Dormancy and Low-Growth States in Microbial Disease

Organisms multiply only when the conditions are beneficial, and when not multiplying, they concentrate on survival of environmental stress. Many bacteria that harm humans survive for most of the period of infection in a low-growth state. This book addresses the basic scientific aspects of microbial dormancy and low-growth states, and places them in the context of human medicine. The book introduces basic scientific aspects of bacterial growth, non-growth, culturability, and viability. Later chapters cover the crucial relationship between low-growth states and survival of stress, the survival of the immune response, and interbacterial signalling. This is followed by chapters on aspects that are of direct importance to medicine, namely antibiotic resistance arising in stationary phase, biofilms, tuberculosis, and the bacteria, which cause gastric ulcers.

ANTHONY R. M. COATES is currently Professor of Medical Microbiology and Chairman of the Department of Medical Microbiology at St. George's Hospital Medical School, London. His research background is in the molecular pathogenesis of infection with a particular interest in tuberculosis.

Published titles

1. *Bacterial Adhesion to Host Tissues.* Edited by Michael Wilson 0521801079
2. *Bacterial Evasion of Host Immune Responses.* Edited by Brian Henderson and Petra Oyston 0521801737

Forthcoming titles in the series

Susceptibility to Infectious Diseases. Edited by Richard Bellamy 0521815258
Mammalian Antimicrobial Peptides. Edited by Deirdre Devine and Robert Hancock 0521822203
The Dynamic Bacterial Genome. Edited by Peter Mullany 0521821576
Bacterial Protein Toxins. Edited by Alistair Lax 052182091X
Bacterial Invasion of Host Cells. Edited by Richard Lamont 0521809541

Over the past decade, the rapid development of an array of techniques in the fields of cellular and molecular biology has transformed whole areas of research across the biological sciences. Microbiology has perhaps been influenced most of all. Our understanding of microbial diversity and evolutionary biology, and of how pathogenic bacteria and viruses interact with their animal and plant hosts at the molecular level, for example, has been revolutionized. Perhaps the most exciting recent advance in microbiology has been the development of the interface discipline of cellular microbiology, a fusion of classic microbiology, microbial molecular biology, and eukaryotic cellular and molecular biology. Cellular microbiology is revealing how pathogenic bacteria interact with host cells in what is turning out to be a complex evolutionary battle of competing gene products. Molecular and cellular biology are no longer discrete subject areas but vital tools and integrated parts of current microbiological research. As part of this revolution in molecular biology, the genomes of a growing number of pathogenic and model bacteria have been fully sequenced, with immense implications for our future understanding of micro-organisms at the molecular level.

Advances in Molecular and Cellular Microbiology is a series edited by researchers active in these exciting and rapidly expanding fields. Each volume will focus on a particular aspect of cellular or molecular microbiology, and will provide an overview of the area, as well as examining current research. This series will enable graduate students and researchers to keep up with the rapidly diversifying literature in current microbiological research.

MCM

Series Editors

Professor Brian Henderson
University College London

Professor Michael Wilson
University College London

Professor Sir Anthony Coates
St. George's Hospital Medical School, London

Professor Michael Curtis
St. Bartholemew's and Royal London Hospital, London

Advances in Molecular and Cellular Microbiology 3

Dormancy and Low-Growth States in Microbial Disease

EDITED BY
Anthony R. M. Coates
St. George's Hospital Medical School, London.

CAMBRIDGE UNIVERSITY PRESS
Cambridge, New York, Melbourne, Madrid, Cape Town, Singapore,
São Paulo, Delhi, Dubai, Tokyo, Mexico City

Cambridge University Press
The Edinburgh Building, Cambridge CB2 8RU, UK

Published in the United States of America by Cambridge University Press, New York

www.cambridge.org
Information on this title: www.cambridge.org/9780521187848

First published 2003
First paperback edition 2011

A catalogue record for this publication is available from the British Library

Library of Congress Cataloguing in Publication data
Dormancy and low-growth states in microbial disease / edited by
    Anthony R. M. Coates.
        p.   cm. – (Advances in molecular and cellular microbiology ; 3)
    Includes bibliographical references and index.
    ISBN 0-521-80940-1 (hardback)
    1. Medical microbiology.   2. Dormancy (Biology)   I. Coates, Anthony R. M.
    II. Series.
    QR46 .D624   2003
    616'.014 – dc21   2002035070

ISBN 978-0-521-80940-5 Hardback
ISBN 978-0-521-18784-8 Paperback

Additional resources for this publication at www.cambridge.org/9780521187848

Contents

The plate facing page 82 is available for download in color from
www.cambridge.org/9780521187848

CONTENTS

Contributors

M. R. Barer
Department of Microbiology and Immunology
Medical Sciences Building
PO Box 138
University Road
Leicester LE1 9HN
UK

Michael R. W. Brown
Department of Pharmacy and Pharmacology
University of Bath
Claverton Down
Bath BA2 7AY
UK

Anthony R. M. Coates
Department of Medical Microbiology
St George's Hospital Medical School
London SW17 0RE
UK

Petra Dersch
Institute für Biologie, Mikrobiologie
Freie Universität Berlin
Königin-Luise-Str. 12-16
14195 Berlin
Germany

C. Stewart Goodwin
Department of Oncology, Gastroenterology
Endocrinology and Metabolism
St. George's Hospital Medical School
London SW17 0RE
UK

Regine Hengge-Aronis
Institute für Biologie, Mikrobiologie
Freie Universität Berlin
Königin-Luise-Str. 12-16
14195 Berlin
Germany

Yanmin Hu
Department of Medical Microbiology
St. George's Hospital Medical School
London SW17 0RE
UK

Gerald C. Johnston
Department of Microbiology and Immunology
Dalhousie University
Halifax Nova Scotia B3H 4H7
Canada

David R. Katz
Department of Immunology and Molecular Pathology
Windeyer Institute, University College London
46 Cleveland Street
London W1T 4JF
UK

Valerie Mizrahi
MRC/NHLS/WITS Molecular Mycobacteriology Research Unit
National Health Laboratory Service
PO Box 1038
Johannesburg 2000
South Africa

Gabriele Pollara
Department of Immunology and Molecular Pathology
Windeyer Institute, University College London
46 Cleveland Street
London W1T 4JF
UK

Hugh W. Pritchard
Seed Conservation Department
Royal Botanic Gardens, Kew
Wellcome Trust Millennium Building
Wakehurst Place
Ardingly
West Sussex RH17 6TN
UK

Richard A. Singer
Department of Biochemistry and Molecular Biology
Dalhousie University
Halifax Nova Scotia B3H 4H7
Canada

Anthony W. Smith
Department of Pharmacy and Pharmacology
University of Bath
Claverton Down
Bath BA2 7AY
UK

Simon Swift
Division of Molecular Medicine and Pathology
Faculty of Medical and Health Sciences
University of Auckland
Private Bag 92019 Auckland
New Zealand

Peter E. Toorop
Seed Conservation Department
Royal Botanic Gardens, Kew
Wellcome Trust Millennium Building
Wakehurst Place
Ardingly
West Sussex RH17 6TN
UK

Digby F. Warner
MRC/NHLS/WITS Molecular Mycobacteriology Research Unit
National Health Laboratory Service
PO Box 1038
Johannesburg 2000
South Africa

CONTRIBUTORS

Preface

All cellular life forms can exist in multiplying, non-multiplying, and slowly multiplying states. In the case of bacteria, slowing of growth is associated with tolerance to a wide range of stresses, such as heat, cold, and antibiotics. This makes sense because organisms multiply only when the conditions are beneficial, and when not multiplying, the organism concentrates on survival in the face of environmental threats.

Many different species of microorganisms live in or on humans. Most are harmless or even beneficial. A few are pathogens. Persistence in a slowly multiplying or non-multiplying form is very common. In some situations, slow multiplication is the dominant phase for bacteria. This book addresses the basic scientific aspects of microbial dormancy and low-growth states and places them in the context of human medicine. An introduction to aspects of growth, non-growth, culturability, and viability is provided in Chapter 1. In the next chapter, the crucial relationship of low-growth states with survival of stress is discussed, with detail on the molecular mechanisms which are involved in the stress response. Chapter 3 sets the scenario for survival of the immune response, and many examples are provided of low-growth states that are associated with different survival mechanisms in animals and in humans. The next two chapters deal with interbacterial signalling in dormancy and mutations in stationary phase, respectively. These are two relatively new areas of study, which are becoming increasingly important. Chapter 6 describes biofilms and their relationship to dormancy and antibiotic resistance. Biofilms are most important in medicine, because many bacterial infections persist as biofilms on, for example, indwelling catheters. Unfortunately, such infections are difficult to eradicate because organisms within the biofilm are tolerant to antibiotics. In the Chapter 7, tuberculosis is discussed. Of all the bacterial infections, this is the classic persistence type, surviving the immune system for decades and requiring six months of chemotherapy due to sub-populations of antibiotic-tolerant bacteria. A similar situation is discussed in Chapter 8, namely gastric ulcers, which are caused by a bacterium which persists in the stomach for years. The final two chapters are on dormancy in eukaryotes. The reason for including these is to show that all cellular life forms have low-growth states which are stress resistant when compared to the multiplying phase. Low growth in yeast is dealt with in Chapter 9, and dormancy in plants in Chapter 10.

Anthony R. M. Coates

CHAPTER 1

Physiological and molecular aspects of growth, non-growth, culturability and viability in bacteria

M. R. Barer

INTRODUCTION

Infection requires growth of pathogens in host tissues or on host epithelia. Cessation of growth is generally correlated with control of infection. Clinically latent infections may reflect microbial growth balanced by host control mechanisms such that the interaction remains below the threshold of detection. Alternatively, the pathogen may have genuinely ceased growth and survive in some form of stasis. In most cases we cannot distinguish between these possibilities. However, there have been important recent advances in our understanding of bacterial populations in which net growth cannot be detected and in recognising the limitations of *in vitro* culture as a means of determining the presence and viability of bacteria. These advances present new opportunities to study the role of non-growing and dormant bacteria in infection and to consider the degree to which culture-based methods may give a false impression of the absence of pathogens during infection, clinical latency and treatment.

The progress of molecular methods in microbiology challenges us to determine the molecular basis of growth and its regulation and to develop such methods to detect growth and viability. In the present context, the long-term aim must be to recognise growth states of microbial populations in the human host.

BACTERIAL GROWTH

Growth involves the accumulation of biomass and may include genomic replication, cell division and an increase in the number of propagules of the organism concerned. For most bacteria it is generally held that, after division, a newly formed cell placed in an environment favourable to growth will double its mass then divide to form two equal-sized progeny via binary fission. This process has been subjected to detailed analysis and is discussed from a highly selective viewpoint here. For more comprehensive and introductory discussions, the reader is referred to recent reviews (35, 55, 57, 66, 82, 104).

Our current understanding of bacterial growth derives overwhelmingly from studying selected organisms in broth cultures. Liquid cultures are convenient; most variables can be precisely controlled, and the scale can be adjusted to provide sufficient biomass for almost any form of analysis. In achieving reproducible results between laboratories, the development of chemically defined media, consistent inocula and the recognition of growth states that can be detected by sequential optical density or turbidity measurements have provided a platform for further development. The widely accepted terminology of lag, exponential (or log) and stationary phases of growth in batch culture provides essential physiological points of reference and these are often applied, with scant justification, to bacterial cells and populations outside the highly defined laboratory environments indicated.

A detailed analysis of the energetics and stoichiometry of bacterial growth has been made possible by analysing bacterial populations growing at constant rates in chemostat or turbidostat cultures (35, 57, 93). These systems provide a relatively reproducible gold standard in which a state referred to as "balanced exponential growth" can be achieved for extended times. The resultant population of cells is generally believed to be uniform and growing at similar rates. Thus it is considered legitimate that analyses of cells in balanced exponential growth can be divided equally amongst all the cells present in the sample to yield estimates of content or activity per cell present.

An important alternative approach has been to start by considering the bacterial cell cycle, which starts with the birth of a cell by binary fission of a parental cell and ends with the division of the new cell. This kind of work draws substantially on our understanding of the eukaryotic cell cycle, where the biochemical and physiological events have been separated into distinct phases (G_1, S, G_2, and M with or without G_0), and has been pursued using techniques that provide large populations of cells that are all at the same stage of the cycle. While some controversy continues, it is generally thought that events that are considered critical for progression through the cell cycle in eukaryotes (e.g., initiation and termination of DNA synthesis) are not similarly regulated in bacteria. Rather, the short-term fate of a cell is determined by the rate at which it accumulates biomass and by the particular size:growth rate ratios at which division is initiated (18). Recently, however, Walker and colleagues (120) have suggested that the *umuDC* component of the bacterial SOS response functions in a manner analogous to the eukaryotic S phase checkpoint. The analogy is complicated by the fact that rapidly growing bacteria initiate new rounds of chromosome synthesis before the last has finished. The authors also point out that the associated checkpoint and DNA repair systems are well suited to dealing with DNA damage accumulated during stationary phase at the time of re-entry into the growth cycle (92).

Most biochemical knowledge obtained with these methods refers to large cell populations ($>10^7$) of readily culturable bacteria in exponential growth phase. Here, we are primarily concerned with the behaviour of pathogens during infections. Not only will these organisms rarely be in a simple suspension phase but also it seems most unlikely that the environment will be conducive to unimpeded exponential growth. Evidently, the degree to which most of our knowledge of bacterial growth is applicable to the environments that primarily concern us must be limited.

Laboratory studies on bacterial growth have also provided limited information regarding growth in colonies on or in solidified laboratory media (74, 125) and in biofilms (37, 75). While information on the growth of bacteria in colonies and in broth may be valuable in designing isolation and culture media for medically important bacteria (35), growth in biofilms is probably a principal mode of bacterial propagation in natural communities. In infections involving fluid-filled spaces (e.g., cystitis) it is plausible that the growth phases recognised in broth culture may be applicable and the relevance of biofilm growth to colonization of intravascular devices also seems certain. However, beyond these examples, assignment of *in vitro*-defined growth phases to pathogens at various stages in infection is largely speculative.

Molecular Information Related to Bacterial Growth

Studies on carefully defined broth cultures remain the principal reliable source of information on the molecular basis of bacterial growth. As key genes involved in growth and its regulation have been identified through recent pre- and post-genomic studies, the possibility of determining the importance of these genes to infection through deletion, over-expression and reporter studies has been extensively exploited. In the context of infection, it is conspicuous that technologies applied to detection of genes essential for growth *in vivo*, such as signature tagged mutagenesis, have often detected genes that appear integral to growth and metabolism (as opposed to classical aspects of virulence such as invasion and toxicity) as essential for *in vivo* survival (e.g., 63, 94).

A somewhat arbitrary selection of genes whose expression has been related to growth in various ways is reviewed below. Ultimately it should be possible to recognise all the genes that are required for growth in specific environments. It seems likely that these will fall into two categories: those required in all circumstances and those required only for special environments.

Ribosomal RNA

A single *E. coli* chromosome generally carries seven copies of the genes encoding ribosomal RNA. In contrast, the *Mycobacterium tuberculosis* chromosome encodes only one copy. Given the greater than tenfold difference in minimum doubling times between these organisms (0.3 h vs. 6 h), it seems likely that this is no accident. The 16S, 23S and 5S genes (and some tRNA genes) are located in tandem and are initially transcribed into RNA as a single molecule, which therefore includes the so-called intergenic transcribed sequences (ITSs). The transcript is then processed into the recognised subunit components, and these combine with ribosomal protein to form functional ribosomes. Aside from the central role now occupied by the 16S molecule and the ITSs in the classifications of *Bacteria* and *Archaea*, the rate at which these genes are transcribed and the 16SrRNA content of bacterial cells has been directly correlated with bacterial growth rates *in vitro* (14, 20). Analysis of these genes and their products in samples therefore presents opportunities to both identify and make some inferences about the protein synthetic capacity and growth rate(s) of the organisms present.

Chromosome Replication

Chromosome replication requires more time to complete than the time available between cell divisions during rapid growth of *E. coli*. The organism circumvents the potential problem of producing cells with less than a single complete genome by initiating rounds of chromosome replication at intervals compatible with the cell replication rate. Initiation always starts at the same locus (*oriC*) and proceeds bi-directionally to the terminus region (76). One consequence of this is that cells in rapidly growing populations contain more than one chromosome replication fork in progress, and the largest cells present (i.e., those close to fission) have a chromosomal DNA content in excess of two copies of the complete genome. The mechanism by which the interval between initiating rounds of chromosomal replication is regulated is not understood, but several gene products are known to be essential. Amongst these the DnaA protein, a DNA-, ATP- and ADP-binding protein, has been most extensively studied and appears to play a central role in assembly of the initiation complex (76). A further consequence of the pattern of replication is suppression of transcription of specific genes as the replication fork passes through. This leads to apparent cell-cycle-related gene regulation in synchronised cultures (129).

Cell Division

Understanding of the molecular basis of bacterial cell division has advanced dramatically over recent years. Progress has been fuelled by development of immunocytochemical techniques for bacteriology and by the use of translational reporter fusions with the green fluorescent protein. These developments have enabled localisation of key molecules that determine the site and process of cell division. Amongst these the tubulin-like molecule FtsZ has been extensively studied. Around 10,000 molecules of this key protein are present in each *E. coli* cell and, like its eukaryotic counterpart, it is present in both soluble and polymerised forms. Location of FtsZ polymers in ring structures indicates the site of prospective septum formation, and using *ftsZ::gfp* translational fusions, it has been possible to observe, in real time, the formation and subsequent contraction of the FtsZ ring in parallel with septum formation and cell division (67, 118). Although FtsZ possesses GTPase activity, it is not known whether it provides the physical force required for septation and fission. Inhibition of FtsZ polymerisation by SulA (a protein produced as part of the SOS response) in growing cells leads to filamentation, thereby illustrating the key role of FtsZ in fission. All bacteria so far studied possess FtsZ homologues, and the relative abundance of the molecule makes it an attractive target for study in clinical samples. The presence of FtsZ rings indicates active cell division, and in *Bacillus subtilis*, asymmetric positioning of the ring indicates the onset of sporulation (64).

Global Regulatory Proteins

These molecules direct differential gene expression by binding either to DNA or to components of the transcription/translation apparatus. Their own levels of expression and activity are modulated by a variety of internal and external stimuli. It would be impractical to discuss even a small minority of these molecules here, but the levels and/or activities of some prominent examples in relation to bacterial growth are outlined in Table 1.1. A full discussion of these molecules is in Chapter 2.

A discussion of the complex regulatory hierarchy and network that are emerging from the study of these proteins and their cognate regulons is beyond the scope of this chapter. The painstaking process of analysing their

Table 1.1. *Regulatory proteins associated with different aspects of growth in bacteria*

Category	Protein	Gene(s)	Some relationships to growth	Function
Regulation via DNA topology[1]	H-NS	osmZ	Levels in constant ratio to DNA content during growth. Depressed in stationary phase	Histone-like DNA binding protein that represses transcription of multiple genes
	LRP	lrp	Repressed by growth in rich medium	Selective repression and activation of genes appropriate to available nutrient sources
	IHF	ihfA (himA) ihfB (himD)	Induction on entry into stationary phase; expression dependent on ppGpp	Interaction with DNA induces 180° bend enabling long-range interactions
SOS response[2]	RecA	recA	Activation of RecA by DNA damage induces cleavage of LexA and de-repression of SOS genes and arrest of cell division.	RecA controlled genes effect DNA repair and maintain λ-like phage lysogeny.
	LexA	lexA		
Alternate σ factors[3] (required for transcription)	σ^{70}	rpoD	σ^S levels increase on entry into stationary phase and on sudden growth arrest.	Main RNA polymerase σ subunit
	σ^{38} (σ^S)	rpoS		Stationary phase and stress induced
	σ^{32}	rpoH		Heat shock induced
	σ^{24} (σ^E)	rpoE	Alternate σ factors appear to compete with σ^{70} for binding to a limited amount of core RNA polymerase. Promoter specificity is modulated by alternate σ factor binding	Induced by extreme heat shock and regulates extracellular proteins
	σ^{28} (σ^F)	fliA		Flagellar gene regulation
	σ^{54}	rpoN(glnF)		

			in combination with many other factors (e.g., H–NS, LRP, IHF & ppGpp). In *B. subtilis* a succession of alternate σ factors directs the programme of gene expression in sporulation.	Control of nitrogen metabolism
Universal stress response proteins[4]	UspA	*uspA*	Induced in late exponential phase and by all known stress responses (σ^{70} and ppGpp dependent)	Regulation via phosphorylation of target proteins?
	UspB	*uspB*	Induced during transition phase (σ^{38} and ppGpp dependent)	Unknown
General metabolic regulators[5]	CRP	*crp*	Low glucose (e.g., on entry into stationary phase) leads to increased cAMP levels.	CRP–cAMP complexes activate or repress specific genes.
	FNR	*fnr (nirA)*	FNR senses oxygen (anaerobic/aerobic growth)	Reduced FNR activates or represses specific genes related to anaerobiosis.
	ArcA	*arcA*	ArcA phosphorylated by cell membrane–associated ArcB quinone-responsive sensor kinase	Regulates catabolism of metabolic reserves

Note: Information presented predominantly relates to studies on *E. coli*. Abbreviations: IHF – integration host factor; USP – universal stress protein; CRP – cAMP receptor protein.

General sources: 78, 82, 83, 106. *Specific sources:* [1](3, 13, 24, 69); [2](120); [3](27, 29, 40, 64, 98, 109); [4](30, 33, 88); [5](11, 31, 34, 52, 62, 87, 105, 107).

respective roles is really only in its early stages and doubtless there are many regulators yet to be recognised and more functions to be defined. The relationships of these molecules to growth suggested in Table 1.1 emerge essentially from studies on samples from populations in specific growth phases during growth in defined media. With a few notable exceptions, information (e.g., 68) on the expression of regulatory proteins in contexts relevant to infection is very limited.

It should not be forgotten that there are many other classes of molecule that regulate bacterial phenotype. The underlying point here is that phenotype and growth state cannot necessarily be inferred from the detection of selective mRNA profiles. At the macromolecular level selective proteases (38) and antisense RNAs (121) have received much attention. Small molecules such as cyclic AMP (cAMP) (11) and guanosine tetraphosphate (ppGpp) (17, 43) are also recognised to have important regulatory roles. Many gene products affect their intracellular levels, and they have plieotropic allosteric effects on their respective binding proteins. The role of ppGpp, the key product of the stringent response, deserves special mention here since, by binding to the B subunit of DNA-dependent RNA polymerase, it provides another means of directing selective gene expression. The stringent response is stimulated by amino acid starvation and is generally associated with growth arrest (17).

In this selective survey the obviously important areas of energy metabolism, cell envelope biosynthesis and assembly and the so-called housekeeping genes have largely been ignored. However, the process of relating the expression of genes to bacterial growth could be extended to cover the entire genome and this can serve little function until we have an adequate interpretive framework. The global approaches offered by proteomics and arrays provide realistic prospects that this will be achieved.

GROWTH AND STASIS

The growth phases of bacteria in batch culture have been reviewed extensively elsewhere (35, 82, 93). Here the focus will be on individual cells and the populations they comprise. The aim is to introduce a framework within which cells in physiological states of particular significance to infection can be recognised and to cross-reference this to the classical growth phases.

Figure 1.1 presents a diagram outlining the various physiological states that can be recognised in relation to the growth of bacteria. Laboratory cultures can be observed at the population or cellular level and a comparison between these is attempted in Fig. 1.1. A central dichotomy is suggested between cells growing or committed to grow and those in some form of non-growing

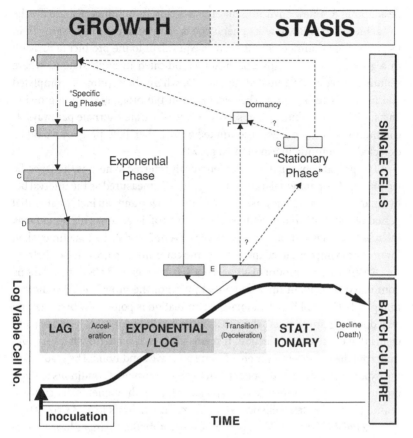

Figure 1.1. Diagram comparing recognised states of growth and stasis in single cells and populations of bacteria. Point A identifies a cell committed to growth and replication; B, the initiation of growth; C, accumulated biomass below that required to initiate septation; D, septation prior to fission. E represents the point at which cells are notionally committed either to continued growth (equivalent to B) or to stasis. The conditions required for commitment to either pathway are only recognised at the population level. G represents cells that are not accumulating biomass, and F represents cells that may be formally described as dormant.

state (stasis). All the states possess potential for further growth and replication, and the cells concerned should therefore be considered viable.

Exponential Phase

In state A, a hypothetical cell, committed to growth and with appropriate resources available but not yet detectably growing, is envisaged. This cell may be adapting to a new environment or recovering from injury. Eventually the cell achieves state B where its phenotype is adapted to commence growth in its current environment and is seen as equivalent to the product of division in a growing culture. This cell grows as indicated through state C, where septum formation is initiated, to state D, where the septum is completed and fission is in progress. The separation of the progeny (E) into growing and static cells is arbitrary and serves only to illustrate alternate pathways. If conditions were conducive to continued growth then both progeny would be expected to continue in exponential growth.

The period between A and B is tentatively referred to here as the "specific lag phase." In operational terms, the lag phase is measured as the interval between inoculation and the onset of detectable growth and can include an initial period of cell death and growth below the limit of detection. This period may include the times indicated between G and A or F and A, i.e., the time taken for non-growing cells to adapt and become committed to growth (see below).

An enormous amount of knowledge has been gained about populations dominated by cells in the exponential phase of the growth cycle indicated in Fig. 1.1. The rapidity of biomass accumulation is potentially breathtaking with doubling times of less than 30 minutes readily achievable by many pathogens that cause acute infection. It seems likely that such growth rates could underpin the rapid development of some infective conditions. The gradient of the exponential phase is dependent on the environmental conditions and the organism. *In vitro*, unrestricted (exponential) growth in chemostat cultures is amenable to quite sophisticated mathematical analysis (e.g., 35, 57, 93).

Bacterial physiological responses to environmental changes have mainly been studied using exponential phase cells. Where these changes are potentially lethal, the responses are referred to as stress responses. The genetic basis for the phenotypic changes elicited by environmental change has been studied extensively, initially by mutational and reporter analysis and at the proteomic level (10, 32, 40, 41, 115, 127). Latterly, genomic and subgenomic arrays have afforded an attractive approach to studying these adaptive responses at a global transcriptional level (119). Depending on the nature of the environmental change or stress, the changes in gene expression elicited may involve between tens and hundreds of different genes. Where the change is not stressful (as defined above) it appears that growth is substantially slowed down or arrested and resumes after the adaptation is complete. In some cases, notably where nutrient depletion precludes further growth, changes in the pattern of gene expression are not confined to a single shift but rather a sequential programme of change is entered into (54). Where this results in a defined morphological adaptive change, such as in sporulation, it is referred to as differentiation (28). In contrast, if the change is stressful (e.g., a substantial pH change or temperature increase) a proportion of the population is killed. The adaptive response in the survivors makes them at least temporarily more resistant (a higher proportion of survivors) to further similar stress.

Stress-responsive genes may be activated by one or more stimuli and in some cases multiple stimuli. Three examples of such multiply responsive genes, *rpoS*, *uspA* and *uspB*, are cited in Table 1.1. It is conspicuous that all three are upregulated in stationary phase. This lends weight to the view that the arrest of net growth in batch culture referred to as stationary phase is itself a form of stress response related to either nutrient depletion or accumulation of toxic metabolites. Several stress responses confer cross-protection against other stresses (e.g., pH and heat) and stationary phase populations are generally more resistant to stress than exponential phase populations (2, 40). (See Chapter 2 for a full discussion of survival of stress.)

Stationary Phase

Although the classical view of stationary phase is that it reflects growth arrest associated with nutrient exhaustion, recent work has raised the possibility that cessation of growth may sometimes be "elective." Indeed there is evidence that bacterial populations have not exhausted all the nutrients they could grow on when they enter stationary phase (7) and that they may produce specific autocrine signals that tell cells not to grow (59). The possible roles for cell-to-cell communication in regulating growth are discussed further below.

Returning to Fig. 1.1, some of the reasons why cells may enter the stasis section of the cycle have been identified. Apart from the ability to reproduce the phenomenon of the stationary phase in batch cultures and to alter the kinetics of its onset by the composition of the growth medium, little is known about how the transition from exponential to stationary phase is regulated. A summary of the recognised influences is given in Fig. 1.2. It is certainly clear that genes such as *uspA*, *uspB* and *rpoS* are upregulated prior to or coincident with the onset of the cessation of net detectable growth (cf. Table 1.1), and it seems likely that they play significant regulatory roles. However, it must be emphasised that, even *in vitro*, it is very difficult to achieve uniform bacterial populations and the standard methods of analysis reflect only gross changes in biomass and dominant biochemical properties. It is therefore entirely possible that, in batch culture, cell populations characterized as being in exponential or stationary phase in fact comprise mixtures of cells in all of the states identified in Fig. 1.1 but in different proportions (e.g.,

11

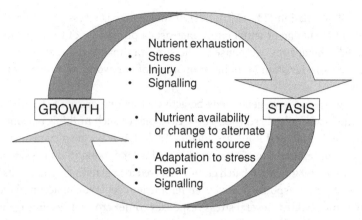

• Nutrient exhaustion
• Stress
• Injury
• Signalling

GROWTH

STASIS

• Nutrient availability or change to alternate nutrient source
• Adaptation to stress
• Repair
• Signalling

Figure 1.2. Diagram illustrating factors influencing growth and stasis at the population level.

100:1 growing:static in exponential phase; 100:1 static:growing in stationary phase). Only in the chemostat or turbidostat, or after multiple rounds of growth and dilution prior to stationary phase, will the static population be kept to a minimum (but still not eliminated).

An important consequence of this last point is that it is very difficult to attribute specific patterns of gene expression to specific cell populations when the measurements have been performed at the population level. Thus, for example, when we observe changes in the patterns of protein expression on 2-D gels in response to a stress, one cannot be certain that individual changes are happening in all the cells sampled. Indeed it is quite possible that multiple subpopulations are represented. Moreover, the relationships between gene expression and phenotype are rarely determined in such experiments. So, where the stress is lethal to a portion of the population, we cannot even be certain whether the changes are taking place in cells that are going to survive or in those that are going to die. These problems are not insurmountable, but they do show some limitations to the global analytical approach.

From the above it should be apparent that while stationary phase can be recognised as a phenomenon and characterised at the molecular level in batch cultures (56, 70, 113), the notion of a "stationary phase cell" is by no means precise. It is well known that smaller cells with lower ribosomal content dominate stationary phase cultures, but similar statements could be made about chemostat cultures at very low dilution rates. In the case of stationary phase resulting from various nutrient limitations, the smaller cells appear to result from reductive cell divisions (fission without cell growth) (54), but it is not certain that this is always the case. One important feature of stationary phase cultures is that they are generally more resistant (in terms of maintaining colony-forming unit (CFU) levels to multiple stresses (e.g., removal of C, N or P from the medium, heat shock or antibiotic treatment). This may well relate to upregulation of genes like *rpoS*, the expression of which is associated with several different stress responses. Teleologically this makes sense since, at least notionally, stationary phase implies a relative lack of resources for the bacteria so it would seem prudent to be protected against multiple noxious influences when the capacity to respond is reduced.

Cells that persist in stasis can be considered to be ageing. When stasis is associated with nutrient exhaustion the capacity for turnover and repair in cells is limited. Recent studies have provided evidence that accumulation of oxidative damage to proteins and DNA is a critical aspect of survival under these conditions (25, 26, 87). These and related studies have been drawn together into a framework for understanding metabolic adaptations to stasis in bacteria that could have far-reaching implications for how we approach the control of non-replicating bacterial populations of concern to medicine and public health (85, 86).

Finally, stationary phase cultures are by no means inert. Stress responses can be detected, and at least in the case of acid stress, a response specific to stationary phase can be demonstrated (32). More significantly, at least in terms of Fig. 1.1, stationary phase cultures are not exclusively composed of non-growing cells. Kolter and colleagues (56) have described a phenomenon referred to as "GASP" (growth adapted to stationary phase) in which long-term stationary phase cultures were shown to contain successive growing subpopulations that replicated from CFU levels below the limit of detection to eventually dominate the CFU population. These emergent populations have been specifically attributed to *rpoS* and *lrp* mutations (128, 130). Thus, while the total CFU count remained the same, this concealed dynamic events occurring within the study population.

Dormancy and Sporulation

In contrast to the notional stationary phase cell shown in Fig. 1.1, there are at least some defined examples of dormant cells. Here, the term dormant is used to denote cells in which there has been a reversible shutdown of metabolic activity (47). The bacterial spore provides the clearest example of a dormant bacterial cell. Sporulation is a differentiation pathway involving sequential activation of genes initially in the mother cell and then selectively in the mother and developing spore cell (27, 116). The process provides a genetic paradigm for differentiation and dormancy in bacteria and has been studied most extensively in *B. subtilis*. In particular, the recognition of the importance of switching between alternate σ factors has provided a useful framework for studying adaptation and differentiation in bacteria. Sporulation leads to the production of highly stress-resistant cells and can be viewed as an extreme form of the adaptations that occur in stationary phase. Indeed, sporulation is initiated in transition phase and by the factors identified in Fig. 1.2. (Note that several peptide signalling factors have been defined in this context (60).) Exactly what decides whether a cell enters stationary phase or dormancy (G or F, Fig. 1.1) is not defined, but it is certainly the case that both spores and vegetative cells are present in stationary phase cultures. This heterogeneous response reinforces the points made above concerning the multiple populations that may be present in bacterial monocultures. (Note that other heterogeneous responses including development of competency and motility also occur during transition phase in *B. subtilis*.)

Somewhat less well defined are the dormant cells of *Micrococcus luteus* described by Kaprelyants, Kell and their colleagues (46, 50). These cells develop slowly after maintenance of stationary phase cultures for several months. In classical terms the decline phase is well established in these cultures since CFU counts have generally fallen by several orders of magnitude by the time the dormant cells can be demonstrated. Although some morphological changes are recognised in the populations containing dormant *M. luteus* cells, Kaprelyants and colleagues explicitly recognise dormant cells by two properties: their substantially reduced capacity to take up rhodamine 123 (a membrane energisation-sensitive fluorescent probe) when compared with exponential or early stationary phase populations and the capacity to be cultured through colony formation or in broth. Through painstakingly careful experiments, these workers were able to demonstrate that dormant cells that could not be cultured by conventional means could nonetheless be resuscitated by exposure to cell-free supernatants from growing *M. luteus* cultures. Subsequently, these supernatants were shown to contain a 17-KD protein that Kaprelyants and colleagues termed resuscitation promoting factor (Rpf) and the cognate gene (*rpf*) was cloned and sequenced (80, 81).

An important distinction must be made between dormancy as exemplified by sporulation and by the *M. luteus* model. Sporulation clearly results from a programme of gene expression that can legitimately be described as differentiation. The *M. luteus* cells meet an operational definition of dormancy (see below) but there is no evidence that they result from a specific genetic programme or that they confer a survival advantage in the way that spores clearly do. Indeed there is evidence that the dormant cells are in fact "injured" or "moribund" since their permeability properties are demonstrably "repaired" during the resuscitation process (44).

Nonetheless, since there is now nucleotide sequence data from many of the genes involved in sporulation and germination in several genera and from *rpf* in *M. luteus*, the opportunity to determine whether other bacteria encode homologues of these genes arises. Sporulation gene homologues have been found in bacteria that have not been demonstrated to produce spores (e.g., 21). However, caution must be exercised in concluding that these organisms have dormant forms that have not been recognised. Several genes that were first recognised in the context of sporulation are now known to have important functions in vegetative cells.

In contrast, whether or not dormant *M. luteus* cells are comparable to spores, studies on purified Rpf have shown that it has distinctive growth-enhancing properties, notably shortening of the lag phase, that have led its discoverers to describe it as the first "bacterial cytokine." *Rpf* homologues appear to be confined to high-GC gram-positive organisms and it is particularly noteworthy that the *Mycobacterium tuberculosis* and *M. leprae* genomes both encode multiple homologues (80).

The decline or death phase of batch cultures is a highly variable phenomenon depending on the organism, strain and medium used. Classically the total cell number in the culture is maintained while the CFU count declines. It should be apparent from the foregoing that the stationary phase is only stationary with respect to total cell counts and net biomass. Moreover, the properties of the culture population become less well defined with increasing time after the end of exponential phase and little of substance can be said about the molecular events that occur after the first few days of this period. However, the recent trend to question whether cells that do not produce colonies on the standard culture medium for the organism concerned (e.g., those developing during the decline phase) may, nonetheless, be considered "viable" makes discussion of this topic more appropriate for the final section of this chapter.

Exit from Dormant or Stationary Cellular States and Re-entry into Growth

Static cells probably do not re-enter the growth cycle simply by reversing the process by which they entered stasis (cf. Fig. 1.1). This is clearly so for spores where germination is clearly not the opposite of sporulation; a comparable process is suggested for other forms of dormancy and stasis. Germination has been demonstrated to depend on one or more specific germination signals, and whether this is so for other forms of static cells is not known (79). Nonetheless, it does seem clear that requirements for initiating growth are somewhat distinct from those necessary for its maintenance.

Defining the nutrients and signals and other conditions necessary for initiating growth is of considerable medical importance. Not only are there several diseases, notably tuberculosis, that have "latent" phases in which it is thought that the pathogen may itself be dormant (4, 39), but also the reliable determination of the presence of organisms in clinical and environmental samples by culture remains absolutely central to patient management and public health monitoring (5). In the former case a complete knowledge of those factors that activate and deactivate growth could enable us to recognise why latent disease reactivates; moreover, we might be able to specifically activate dormant cells to make them susceptible to standard chemotherapy. Regarding cultivation, only when we have a comprehensive understanding of those factors necessary for organisms to initiate growth can we be confident that culture-based detection is at its most sensitive.

Apart from germination, we really know very little about the process of transition from stasis to growth. The discovery of Rpf provides exciting opportunities to study one example of this process particularly because a key assayable effect of this molecule is a reduction in the lag phase. Kaprelyants and Kell have reviewed the evidence relating to other molecules that have effects comparable to Rpf (48). Growth stimulatory molecules affecting both Gram-positive and Gram-negative organisms have been identified with varying degrees of certainty.

It seems likely that signalling molecules play a significant role in growth regulation in at least some species in some environments. It has even been suggested that growth of some species may be completely dependent on molecules like Rpf (48). While this is an interesting proposal, demonstration of Rpf dependency requires very specific environmental conditions and the natural physiological roles for the molecule and its homologues are far from being understood. Whatever these may be, the significance of signalling in bacterial growth cannot be ignored; in particular the possibility that growth is a "social," "communal" or "quorum-dependent" process in bacteria is now firmly on the agenda (see Chapter 4).

CULTURABILITY AND VIABILITY

Molecular analyses have allowed us to test for the presence of specific organisms without demonstrating their capacity to multiply. Such developments challenge us to review the value of culture as a means both of detecting the microbes present and determining their viability. In the present context there are three central questions:

1. How representative is culture as a means of detecting all the microbial cells present that are relevant to infection?
2. Are there physiological conditions that might affect the culturability of such cells?
3. Can viability be determined independently of culturability?

Discussion of the problems that beset terminology in this area will not be repeated here (4–6, 44, 51). In order that the reader should at least recognise the framework within which the problematic terms are used here, a table of definitions is provided (Table 1.2). Clear distinctions are drawn between use of terms in conceptual (theoretical) and operational (practical) contexts.

Culturability

Mycobacterium leprae and *Treponema pallidum* stand as monuments to the fact that some known pathogens still cannot be propagated in axenic cultures a century after they were first described. On a larger scale, it has been recognised by microbial ecologists that, when comparisons are made between the total number of bacterial cells present in almost any environmental sample and the total number of colony-forming units obtained by non-selective agar culture, the former exceeded the latter by anywhere from one to three orders of magnitude. This excess of cells that fail to form colonies has been referred to as the "great plate count anomaly" (114).

Using methods based on direct recovery of 16S rRNA subunit genes from samples (36) and *in situ* hybridisation to ribosomes at the single cell level (1), it has been possible to specifically recognise and classify many bacteria that have not been recovered or characterized in laboratory cultures. Of the rDNA sequences recovered in studies of this sort, the majority are apparently derived from organisms that have not previously been characterized. The detection of such "as yet uncultivated" (AYU) bacteria has fuelled the view that culture methods do not adequately represent the range or numbers of bacteria present in most samples containing a diverse microbial flora.

In the context of infection, studies on Whipple's disease and the description of *Tropheryma whippelii* on the basis of 16S rRNA sequence data have clearly demonstrated the potential of molecular approaches to resolve some longstanding problems (99, 101, 123). Taken together with the studies on bacillary angiomatosis and the description of *Mycobacterium genavense*

Table 1.2. *Definitions of key terms relating to growth, culturability and viability of bacteria*

Term	Conceptual definition	Operational definition
Viable	Retaining the capacity for replication over a stated or generally accepted time frame	Explicit demonstration of replication in a validated laboratory system
Replication	Genomic replication and segregation into a new self-propagating unit (propagule)	Observed cell fission or increase in number of propagules
Culturable	Capable of detectable replication in a realisable laboratory system	Detected replication in a validated laboratory system
Growth	Accumulation of biomass	Demonstrated accumulation of biomass
Dormancy	A reversible state of low metabolic activity in a unit that maintains viability	A demonstrated reversible low state of metabolic activity demonstrated by a specific technique or set of techniques in an operationally viable unit
Resuscitation	Transition from a temporary state in which the specified unit had lost the capacity for self-replication to regain that capacity	A demonstrated transition from a temporary state of nonculturability in a defined system to culturability in that system; procedure must exclude regrowth as a possible explanation
Regrowth	Growth and/or replication within a population from below to above to levels of propagules that can be detected by a defined culture test	Growth and/or replication demonstrated by one method that led to detection of the organism by another, less sensitive, method; If latter were applied alone, resuscitation would appear to have occurred

Term		
Cryptic growth	Growth and/or replication of a subpopulation within the study population that cannot be detected by the methods applied	Growth and/or replication of a subpopulation demonstrated by one method that was not detected by another, less sensitive, method; total cell number is not measurably affected
Suicide	An irreversible process by which a viable unit determines the loss of its own viability by a specific mechanism	A non-reversible response in which a population of propagules can be shown to have lost demonstrable viability as a result of a process intrinsic to that response
Activity	A metabolic or behavioural process occurring within the unit under consideration; unit may be viable or non-viable	A demonstrated behavioural, biochemical or physiological process occurring within the unit under consideration
Survival	Maintenance of viability	Maintenance of operational viability

(12, 100), the *T. whippelii* story would seem to endorse the view that molecular approaches could allow us to work with organisms that permanently defy our attempts at cultivation. However, it should be noted that both *M. genavense* and *T. whippelii* have now been propagated *in vitro*, though not axenically in the case of the latter (19, 108).

In any given study the cells that are not recovered in culture potentially comprise two populations: those belonging to species for which no culture method has yet been devised and cells of organisms that can normally be cultured by the method applied but which cannot be recovered for one reason or another. These latter "non-recoverable" cells may include dead, moribund, injured and other "temporarily non-culturable" cells and are discussed more extensively below.

In the present context it is of particular concern that bacteria involved in latent or non-patent infections may be in physiological states that make them resistant to conventional *in vitro* culture. It appears plausible that dormant or non-replicating cells might be difficult to recover in culture. However, this is not generally the case for spores, which can be readily recovered. Instances in which temporarily non-culturable bacterial cells can apparently be recovered are discussed elsewhere in this volume. The physiological bases for these phenomena remain obscure. If differentiation processes are involved then they have yet to be elucidated, and the degree to which the controversial concept of a "viable but non-culturable" (VBNC) state provides a useful physiological framework is discussed in the next section.

For the present discussion three underlying reasons can be recognised for failing to cultivate cells of a normally culturable bacterium:

1. Competition with other organisms: growth of the organism is suppressed or obscured by other organisms present in the same environment.
2. Sub-lethal Injury: the organism is in the process of recovering from some form of damage (e.g., the effect of an antibiotic). During the recovery period the organisms may not be recoverable on conventional isolation media, but if they are provided with appropriate conditions, they can regain "culturability."
3. Lethally injured, moribund and "dead" cells: such cells are non-viable (see below) but they may still be detected by molecular or microscopic methods.

Cells in categories 2 and 3 are readily produced in laboratory experiments. However, their explicit recognition and significance in natural samples remains problematic. In classical operational terms, sub-lethally injured cells fail to grow under certain selective isolation conditions but can still grow under non-selective conditions. The framework for these effects is well developed in food microbiology. Overall, the effect of injury detected this way is to render the injured cells more fastidious in their growth requirements; it also appears that injured cells incubated under inappropriate conditions do not simply fail to grow, they actually lose viability.

While this approach provides a practical means of recognising injury, it is clear that patterns of damage and repair can occur that do not fit into this pattern. For example DNA damage that elicits the SOS response causes an arrest in DNA replication and cell division while SOS-mediated repair continues (120). This process is not detected by simple colony counting measures since growth is only delayed, not prevented. Nonetheless, this process provides a second paradigm for injury and repair that is perhaps more familiar from an anthropomorphic perspective. Cells affected by some inimical stimulus (e.g., an antibiotic or an immune effector system) acquire damaged components that must be replaced by a repair process. During the repair process conditions should be optimised for turnover and replacement of damaged components rather than growth. This conceptual framework has led to the development of "recovery" media specifically to fulfil this role.

There are no satisfactory generalisations that can be made about the recovery of injured bacteria. What matters from a medical perspective is whether they retain pathogenic potential and, if so, whether this is altered compared to "healthy" cells and how the injured cells may be detected. At the time of writing the available evidence suggests that the pathogenic potential of sub-lethally injured cells is reduced (16, 61, 111). Optimal recovery conditions appear specific to the nature of the injury. In one study the culture conditions giving the highest yield from starved cultures was shown to vary almost day by day (81).

In addition to the temporary effects of sub-lethal injury on culturability, a number of other phenomena have clear potential to affect the outcome of a culture assay. These include the ageing process referred to above (effectively oxidatively injured cells), the fascinating phenomenon of "substrate-accelerated death" and other forms of metabolic self-destruction, the potential activation of toxin–antitoxin systems or lysogenic phages by the conditions of the culture test and finally the influence of bacterial cytokines and other intercellular signals (6).

The third group of non-culturable cells identified above has clearly lost all ability to initiate new infections. However these cells may retain some metabolic capacity, which, in special circumstances, may include toxin production and release (53, 97); they may also possess endotoxins.

A general problem with classifying cells into these three groups is that assignment can only be made in retrospect after recovery by *in vitro* cultivation has succeeded or failed (see operational definitions in Table 1.2 and discussions in 6 and 51). Moreover, the assignments can only be crude estimates since the technical difficulties involved in determining the correlation between cells counted by microscopy or molecular methods and CFU are very substantial. In consequence, we can only speculate about the contributions made by such cells to latent or non-patent infections. Nonetheless, it is clear that currently available molecular detection methods make none of the distinctions indicated. Moreover, with some notable exceptions (23, 110), such methods do not currently distinguish between viable and non-viable organisms.

Viability

Viability determinations have played a central role in bacteriology (6, 95, 96). In medical microbiology we are primarily concerned with measuring the effects of antimicrobial agents, sterilisation and disinfection regimens and immune effector mechanisms. In addition, only viable bacteria are considered capable of transmitting infection. For many years it has been accepted practice to equate viability with culture-based estimations such as CFU or MPN (most probable number) counts (MPN counts are a well standardised limiting dilution method (15, 22, 65)). However, partly fuelled by some of the issues raised in the previous section, a number of workers have pointed out that culture may not be adequate to the task (72, 90, 103). Moreover, it has been suggested that some methods other than culture (sometimes termed indirect methods (95)) may better reflect the distribution and activity of bacteria (89). The resultant discussion has often focussed on the suggestion that many bacteria may enter a state referred to as "viable but non-culturable (VBNC)" (4, 5, 44, 51, 126).

The discussion must start with a definition of viability (Table 1.2). Established precedent indicates that a viable entity has the capacity to persist into the future and to carry out those functions with which it is normally associated (6, 51, 95). Because the individual bacterial life span is essentially the time between cell divisions and because this is a variable that depends on growth state, it has been generally accepted that the capacity for replication is implicit in viability. This capacity is unambiguously demonstrated in CFU and MPN tests. However, dependence on culture tests for viability determinations leaves us devoid of a means of assessing viability in AYU bacteria and imposes time delays on analytical work. These problems have provided incentives to further develop indirect methods.

Previous discussions of indirect techniques have concentrated on divisions between different technical approaches. Here, a classification system based on the different aspects of cellular and community function is presented (Table 1.3). The issues relating to these assays have been reviewed elsewhere (6, 51). To summarise, it is possible both in theory and in practice to recognise circumstances in which any of these tests fail to identify viable or non-viable cells as defined above. Nonetheless, very good correlations can be obtained between the indirect test results and culture tests in specific settings (e.g., 42, 45). Problems arise when the context in which the test has been applied is altered. In particular, it cannot be assumed that correlations will be as good when the assay is applied to different organisms or with different inimical stimuli where the mechanism of cell death may be different; re-validation is desirable with respect to every change in conditions.

In spite of these problems, indirect tests can be very useful. In particular, the recognition of trends through quantitative analysis of serial samples using a reproducible method can be informative. Moreover, many problems fall away if we ignore the potential relationships of test results to viability and simply report them as tests of function with all the usual range of considerations regarding confounding factors and the need for appropriate controls (73). That such test results have some relationship to viability is not denied, but the relationship is variable and justifies no more than rather vague statements about the probability that a given cell is viable or that a given culture is more or less viable than a comparable culture. Of course, the statements can be more precise when the conditions are closer to those in which the test was validated. However, in one well studied *in vitro* model, the best that could be achieved was a likelihood that a positive test of cytoplasmic membrane energisation indicated a 19% probability that a given cell could be recovered in culture (49).

A central conclusion from the foregoing is that culturability currently remains the only *unambiguous* operational means of demonstrating viability. This clearly presents a problem for AYU bacteria and the only rigorously defendable view is that, for the most part, the viability of cells and populations belonging to AYU taxa must be considered indeterminate. There are two riders to this view. Firstly, the viability of *M. leprae* can clearly be recognised by propagation of the organism in a suitable animal host. Secondly, it is clear that when cells of an organism are detected in a sample, they must have come from somewhere and they must have been the progeny of a viable cell. Depending

Table 1.3. *A functional classification of indirect tests that have been applied to assess the viability of bacteria*

Function assayed	General method	Examples
Cell "integrity"	Quantitation of cells	Optical density, nephelometry, flow cytometry, Coulter counting
	Microscopy	Morphology of cell by light and electron microscopy; change in phase contrast appearance
Cell permeability	Dye exclusion	Propidium iodide exclusion
	Retention of intracellular components	Assay of released DNA or enzyme (e.g., lactate dehydrogenase)
Cell nucleic acid content	Cell-associated DNA and RNA	DAPI staining of DNA; rRNA *in situ* hybridisation
Cell enzyme content	Modification of chromogenic or fluorogenic substrate	Fluorescein diacetate hydrolysis; tetrazolium reduction
Energy status	Presence of property that requires constant energy input	ATP content, ATP/ADP ratio; rhodamine 123 accumulation (membrane energisation)
Evidence of integrated function	Assay of property that requires function of multiple cell systems	Reporter gene expression (e.g., *luxAB*); inducible reporter gene expression (e.g., *lacZ*); labelled precursor incorporation into specific cell product via multiple processing steps (amino acid incorporation, CO_2 production, ^{3}H-tymidine incorporation)
Transcription*	Detection of mRNA	Uracil incorporation; specific transcript detection by hybridisation or RT–PCR
Evidence of growth potential	Detection of properties that always occur in growing cells	Presence of septa or multiple nucleoids; cell elongation (Kogure test)

Note: Specific references to these methods may be found in (6, 51). RTPCR = reverse-transcription polymerase chain reaction.

on the dynamics of the system sampled, the necessary proximity of a parental cell can be inferred. In many systems (including the human body) the likelihood of engulfment and destruction by a eukaryotic cell is high. Thus, presence of a cell in a high-turnover system implies that either the cell is viable, that viable "relatives" are close by or that the cell is resistant to destruction.

Ultimately it must be our aim to produce validated tests that explicitly demonstrate bacterial replication in natural environments. To some extent, it has been possible to do this with culturable bacteria by recognising the dilution of replication of non-replicating markers (71, 112); however the technical challenge of achieving this with AYU bacteria remains substantial.

The VBNC Controversy

Against the background outlined above, a substantial body of published work has accumulated over the past 20 years relating to bacterial cells that are described as VBNC or in a VBNC state. Some authors have applied this term to AYU bacteria. Although semantically there must be VBNC bacteria out there that belong to AYU taxa, as noted above, the viable cells cannot be explicitly identified. Moreover, the area of controversy is concentrated on operationally non-culturable cells of culturable taxa, and adherents to the VBNC terminology predominantly apply the phrase in this context. The key issues have been reviewed extensively elsewhere and a brief summary is provided below (5, 6, 51, 72, 89, 103).

Colwell and associates first applied the phrase VBNC to cultures of *Vibrio cholerae* that retained some form of demonstrable cellular activity but which failed to grow in standard culture tests for this organism (126). The work had been stimulated by attempts to define the environmental reservoir for this organism. Indeed the possibility of persistence of pathogens in environmental reservoirs in non-culturable forms remains an attractive hypothesis for any bacterial infection in which the distribution of cases is not concordant with the potential sources and reservoirs of the causal organism as identified by culture (5). Similarly, it has been tempting to suggest that related phenomena might be responsible for non-patent periods of infections such as tuberculosis (4).

While the VBNC hypothesis has proved highly stimulating in relation to expanding our views of bacterial physiology and the transmission of infection, both the degree to which putative VBNC cells can be recognised as separate from any of the categories identified in the previous section and the general value of the terminology are contentious.

Non-culturable bacterial cells have been described as VBNC on the basis of retained cellular integrity or activity and by demonstration of their return to culturability. The latter has been achieved by special recovery methods or by animal passage (84, 91, 102, 122). At one level it is clear that some of the reported phenomena can be attributed to injured cells that were temporarily non-culturable by the standard method. The recent demonstration of peroxide-sensitive cells as an explanation for "VBNC phenomena" in *Vibrio vulnificus* provides a prime example of this (9). Further, if cells are identified as VBNC by ultimately returning them to culture then they were never non-culturable in the first place (only temporarily so). Finally, it should also be noted that it can be very difficult to recognise transition from an authentic temporarily non-culturable state to a culturable state and that the experimental protocols of several published studies can be criticised on this basis (51). Thus the validity of the VBNC terminology can be questioned on the basis of established terminology, semantics and experimental technique.

The key issue for infection remains the mechanism by which cells can become non-culturable. The VBNC debate has raised the possibility that some bacteria can give rise to temporarily non-culturable forms effectively by differentiation rather than by injury or death. If this does happen we are faced with the possibility that culture tests do not accurately reflect the destructive effects of antibiotics and immune effector mechanisms. While many intriguing phenomena have been described in this area, the available evidence does not justify such concern. Moreover, no genetic programme giving rise to temporarily non-culturable cells that have long-term survival potential has been specifically identified (87).

It seems likely that phenomena that have been considered to reflect VBNC cells are transient and result from cell injury and degeneration rather than adaptation or differentiation.

CONCLUSIONS

Recent developments in understanding the physiological basis of bacterial growth and survival in the stationary phase have opened up many possibilities to recognise the importance to infection of genes involved in these processes. There is very little information concerning the roles these genes may play during active and latent infection and this must surely be a fertile area for further study.

The detection of many as-yet-uncultured or highly fastidious bacteria that associate with man by molecular methods (8, 58, 77, 117, 124) serves to reinforce recognition of the weaknesses of culture as a means of defining the distribution of bacteria. It also provides a major incentive to develop robust molecular methodologies to characterize the physiological states of these organisms *in vivo*.

The degree to which non-culturable forms of culturable pathogens make a significant contribution to transmission, pathogenesis and latency of infection remains controversial. At present the balance of evidence seems to favour the view that non-culturable forms of such organisms are generally sub-lethally injured or moribund and that they do not result from a specific programme of gene expression.

REFERENCES

1. Amann, R. I., Ludwig, W., and Schleifer, K. H. 1995. Phylogenetic identification and in situ detection of individual microbial cells without cultivation. Microbiol. Rev. **59**:143–169.
2. Antelmann, H., Engelmann, S., Schmid, R., and Hecker, M. 1996. General and oxidative stress responses in *Bacillus subtilis*: Cloning, expression, and mutation of the alkyl hydroperoxide reductase operon. J. Bacteriol. **178**:6571–6578.
3. Aviv, M., Giladi, H., Schreiber, G., Oppenheim, A. B., and Glaser, G. 1994. Expression of the genes coding for the *Escherichia coli* integration host factor are controlled by growth phase, rpoS, ppGpp and by autoregulation. Mol. Microbiol. **14**:1021–1031.
4. Barer, M. R. 1997. Viable but non-culturable and dormant bacteria: Time to resolve an oxymoron and a misnomer? J. Med. Microbiol. **46**:629–631.
5. Barer, M. R., Gribbon, L. T., Harwood, C. R., and Nwoguh, C. E. 1993. The viable but non-culturable hypothesis and medical microbiology. Rev. Med. Microbiol. **4**:183–191.
6. Barer, M. R., and Harwood, C. R. 1999. Bacterial viability and culturability. Adv. Microbial Physiol. **41**:93–137.
7. Barrow, P. A., Lovell, M. A., and Barber, L. Z. 1996. Growth suppression in early-stationary-phase nutrient broth cultures of *Salmonella typhimurium* and *Escherichia coli* is genus specific and not regulated by sigma S. J. Bacteriol. **178**:3072–3076.
8. Beswich, A. J., Lawley, B., Fraise, A. P., Pahor, A. L., and Brown, N. L. 1999. Detection of *Alloiococcus otitis* in mixed bacterial populations from middle-ear effusions of patients with otitis media. Lancet **354**:386–389.
9. Bogosian, G., and Bourneuf, E. V. 2001. A matter of bacterial life and death. EMBO Rep. **2**(9):770–774.
10. Booth, I. R., and Louis, P. 1999. Managing hypoosmotic stress: Aquaporins and mechanosensitive channels in *Escherichia coli*. Curr. Opin. Microbiol. **2**:166–169.
11. Botsford, J. L., and Harman, J. G. 1992. Cyclic AMP in prokaryotes. Microbiol. Rev. **56**:100–122.

12. Bottger, E. C., Teske, A., Kirschner, P., Bost, S., Chang, H. R., Beer, V., and Hirschel, B. 1992. Disseminated *Mycobacterium genavense* infection in patients with AIDS. Lancet **340**:76–80.
13. Bouvier, J., Gordia, S., Kampmann, G., Lange, R., Hengge-Aronis, R., and Gutierrez, C. 1998. Interplay between global regulators of *Escherichia coli*: Effect of RpoS, Lrp and H-NS on transcription of the gene *osmC*. Mol. Microbiol. **28**:971–980.
14. Bremer, H., and Dennis, P. P. 1987. Modulation of chemical composition and other parameters of the cell by growth rate. In Escherichia coli *and* Salmonella typhimurium: *Cellular and Molecular Biology*, pp. 1527–1542 (F. C. Neidhardt, J. L. Ingraham, K. B. Lowet *et al.*, Eds.). ASM Press, Washington, D.C.
15. Button, D. K., Schut, F., Quang, P., Martin, R., and Robertson, B. R. 1993. Viability and isolation of marine bacteria by dilution culture theory, procedures, and initial results. Appl. Environ. Microbiol. **59**:881–891.
16. Caro, A., Got, P., Lesne, J., Binard, S., and Baleux, B. 1999. Viability and virulence of experimentally stressed nonculturable *Salmonella typhimurium*. Appl. Environ. Microbiol. **65**:3229–3232.
17. Cashel, M., Gentry, D. M., Hernandez, V. J., and Vinella, D. 1996. The stringent response. In Escherichia coli *and* Salmonella: *Cellular and Molecular Biology*, pp. 1458–1496 (F. C. Neidhardt, Ed.). ASM Press, Washington, D.C.
18. Cooper, S. 1991. *Bacterial Growth and Division: Biochemistry and Regulation of Prokaryotic and Eukaryotic Division Cycles*. Academic Press, San Diego.
19. Coyle, M. B., Carlson, L. C., Wallis, C. K., Leonard, R. B., Raisys, V. A., Kilburn, J. O., Samadpour, M., and Bottger, E. C. 1992. Laboratory aspects of *Mycobacterium genavense*, a proposed species isolated from AIDS patients. J. Clin. Microbiol. **30**:3206–3212.
20. Delong, E. F., Wickham, G. S., and Pace, N. R. 1989. Phylogenetic stains – ribosomal RNA-based probes for the identification of single cells. Science **243**:1360–1363.
21. DeMaio, J., Zhang, Y., Ko, C., Young, D. B., and Bishai, W. R. 1996. A stationary-phase stress-response sigma factor from *Mycobacterium tuberculosis*. Proc. Natl. Acad. Sci. USA **93**:2790–2794.
22. Department of Health and Social Security, W. O., Ministry of Housing and Local Government 1969. *The Bacteriological Examination of Drinking Water Supplies*. Her Majesty's Stationery Office, London.
23. Desjardin, L. E., Perkins, M. D., Wolski, K., Haun, S., Teixeira, L., Chen, Y., Johnson, J. L., Ellner, J. J., Dietze, R., Bates, J., *et al.* 1999. Measurement of sputum *Mycobacterium tuberculosis* messenger RNA as a surrogate for response to chemotherapy. Am. J. Respir. Crit. Care Med. **160**:203–210.

24. Dorman, C. J., Hinton, J. C. D., and Free, A. 1999. Domain organization and oligomerization among H-NS-like nucleoid-associated proteins in bacteria. Trends. Microbiol. **7**:124–128.

25. Dukan, S., and Nystrom, T. 1998. Bacterial senescence: Stasis results in increased and differential oxidation of cytoplasmic proteins leading to developmental induction of the heat shock regulon. Genes Dev. **12**:3431–3441.

26. Dukan, S., and Nystrom, T. 1999. Oxidative stress defense and deterioration of growth-arrested *Escherichia coli* cells. J. Biol. Chem. **274**:26,027–26,032.

27. Errington, J. 1993. *Bacillus subtilis* sporulation: Regulation of gene expression and control of morphogenesis. *Microbiol. Rev.* **57**:1–33.

28. Errington, J. 1996. Determination of cell fate in *Bacillus subtilis*. Trends. Genet. **12**:31–34.

29. Farewell, A., Kvint, K., and Nystrom, T. 1998. Negative regulation by RpoS: A case of sigma factor competition. Mol. Microbiol. **29**:1039–1051.

30. Farewell, A., Kvint, K., and Nystrom, T. 1998. *uspB*, a new sigmaS-regulated gene in *Escherichia coli* which is required for stationary-phase resistance to ethanol. J. Bacteriol. **180**:6140–6147.

31. Ferenci, T. 1999. Regulation by nutrient limitation. Curr. Opin. Microbiol. **2**:208–213.

32. Foster, J. W. 1999. When protons attack: Microbial strategies of acid adaptation. Curr. Opin. Microbiol. **2**:170–174.

33. Freestone, P., Nystrom, T., Trinei, M., and Norris, V. 1997. The universal stress protein, UspA, of *Escherichia coli* is phosphorylated in response to stasis. J. Mol. Biol. **274**:318–324.

34. Georgellis, D., Kwon, O., and Lin, E. C. 2001. Quinones as the redox signal for the arc two-component system of bacteria. Science **292**(5525):2314–2316.

35. Gerhardt, P. 1994. Section II: Growth. In *Methods for General and Molecular Bacteriology*, pp. 135–294 (P. Gerhardt, R. G. E. Murray, W. A. Wood, and N. R. Krieg, Eds.). American Society for Microbiology, Washington, D.C.

36. Giovannoni, S. J., Britschgi, T. B., Moyer, C. L., and Field, K. G. 1990. Genetic diversity in Sargasso Sea bacterioplankton. Nature **345**:60–63.

37. Gottenbos, B., vanderMei, H. C., and Busscher, H. J. 1999. Models for studying initial adhesion and surface growth in biofilm formation on surfaces. Meth. Enzymol. **310**:523–534.

38. Gottesman, S. 1999. Regulation by proteolysis: Developmental switches. Curr. Opin. Microbiol. **2**:142–147.

39. Grange, J. M. 1992. The mystery of the mycobacterial 'persistor.' Tuberc. Lung Dis. **73**:249–251.

40. Hengge-Aronis, R. 1999. Interplay of global regulators and cell physiology in the general stress response of *Escherichia coli*. Curr. Opin. Microbiol. **2**:148–152.

41. Ishihama, A. 1997. Adaptation of gene expression in stationary phase bacteria. Curr. Opin. Genet. Dev. **7**:582–588.
42. Jepras, R. I., Carter, J., Pearson, S. C., Paul, F. E., and Wilkinson, M. J. 1995. Development of a robust flow cytometric assay for determining numbers of viable bacteria. Appl. Environ. Microbiol. **61**:2696–2701.
43. JoseleauPetit, D., Vinella, D., and Dari, R. 1999. Metabolic alarms and cell division in *Escherichia coli*. J. Bacteriol. **181**:9–14.
44. Kaprelyants, A. S., Gottschal, J. C., and Kell, D. B. 1993. Dormancy in nonsporulating bacteria. FEMS Microbiol. Rev. **104**:271–286.
45. Kaprelyants, A. S., and Kell, D. B. 1992. Rapid assessment of bacterial viability and vitality by rhodamine 123 and flow cytometry. J. Appl. Bact. **72**:410–422.
46. Kaprelyants, A. S., and Kell, D. B. 1993. Dormancy in stationary-phase cultures of *Micrococcus luteus* – flow cytometric analysis of starvation and resuscitation. Appl. Environ. Microbiol. **59**:3187–3196.
47. Kaprelyants, A. S., and Kell, D. B. 1993. The use of 5-cyano-2,3-ditolyl tetrazolium chloride and flow-cytometry for the visualization of respiratory activity in individual cells of *Micrococcus luteus*. J. Microbiol. Meth. **17**: 115–122.
48. Kaprelyants, A. S., and Kell, D. B. 1996. Do bacteria need to communicate with each other for growth? Trends Microbiol. **4**:237–242.
49. Kaprelyants, A. S., Mukamolova, G. V., Davey, H. M., and Kell, D. B. 1996. Quantitative analysis of the physiological heterogeneity within starved cultures of *Micrococcus luteus* by flow cytometry and cell sorting. Appl. Environ. Microbiol. **62**:1311–1316.
50. Kaprelyants, A. S., Mukamolova, G. V., and Kell, D. B. 1994. Estimation of dormant *Micrococcus luteus* cells by penicillin lysis and by resuscitation in cell free spent culture medium at high dilution. FEMS Microbiol. Lett. **115**:347–352.
51. Kell, D. B., Kaprelyants, A. S., Weichart, D. H., Harwood, C. R., and Barer, M. R. 1998. Viability and activity in readily culturable bacteria: A review and discussion of the practical issues. Antonie Van Leeuwenhoek **73**:169–187.
52. Kiley, P. J., and Beinert, H. 1998. Oxygen sensing by the global regulator, FNR: The role of the iron–sulfur cluster. FEMS Microbiol. Rev. **22**:341–352.
53. Kimmitt, P. T., Harwood, C. R., and Barer, M. R. 1999. Induction of type 2 shiga toxin synthesis in *Escherichia coli* O157 by 4-quinolones. Lancet **353**: 1588–1589.
54. Kjelleberg, S., Albertson, N., Flardh, K., Holmquist, L., Jouper-Jaan, A., Marouga, R., Ostling, J., Svenblad, B., and Weichart, D. 1993. How do

non-differentiating bacteria adapt to starvation? Antonie van Leeuwenhoek 63:333–341.

55. Koch, A. L. 1997. Microbial physiology and ecology of slow growth. Microbiol. Mol. Biol. Rev. 61:305.

56. Kolter, R., Siegele, D. A., and Tormo, A. 1993. The stationary phase of the bacterial life cycle. Annu. Rev. Microbiol. 47:855–874.

57. Kovarova-Kovar, K., and Egli, T. 1998. Growth kinetics of suspended microbial cells: From single-substrate-controlled growth to mixed-substrate kinetics. Microbiol. Mol. Biol. Rev. 62:646–666.

58. Kroes, I., Lepp, P. W., and Relman, D. A., 1999. Bacterial diversity within the human subgingival crevice. Proc. Natl. Acad. Sci. USA 96:14,547–14,552.

59. Lazazzera, B. A., 2000. Quorum sensing and starvation: Signals for entry into stationary phase. Curr. Opin. Microbiol. 3:177–182.

60. Lazazzera, B. A., and Grossman, A. D. 1998. The ins and outs of peptide signaling. Trends Microbiol. 6:288–294.

61. LeChevallier, M. W., Singh, A., Schiemann, D. A., and McFeters, G. A. 1985. Changes in virulence of waterborne enteropathogens with chlorine injury. Appl. Environ. Microbiol. 50:412–419.

62. Lee, Y. S., Han, J. S., Jeon, Y., and Hwang, D. S. 2001. The arc two-component signal transduction system inhibits in vitro Escherichia coli chromosomal initiation. J. Biol. Chem. 276(13):9917–9923.

63. Lestrate, P., Delrue, R. M., Danese, I., Didembourg, C., Taminiau, B., Mertens, P., De Bolle, X., Tibor, A., Tang, C. M., and Letesson, J. J. 2000. Identification and characterization of in vivo attenuated mutants of Brucella melitensis. Mol. Microbiol. 38(3):543–551.

64. Levin, P. A., and Grossman, A. D. 1998. Cell cycle and sporulation in Bacillus subtilis. Curr. Opin. Microbiol. 1:630–635.

65. Mackie, T. J., McCartney, J. E., and Collee, J. G. 1996. Mackie & McCartney Practical Medical Microbiology. Churchill Livingstone, London.

66. Madigan, M. T., Martinko, J. M., and Parker, J. 2000. Microbial growth. In Brock Biology of Microorganisms, pp. 740–772. Prentice-Hall, Upper Saddle River, New Jersey.

67. Margolin, W. 1998. A green light for the bacterial cytoskeleton. Trends Microbiol. 6:233–238.

68. Marshall, D. G., Bowe, F., Hale, C., Dougan, G., and Dorman, C. J. 2000. DNA topology and adaptation of Salmonella typhimurium to an intracellular environment. Philos. Trans. R. Soc. Lond. B Biol. Sci. 355:565–574.

69. Marshall, D. G., Sheehan, B. J., and Dorman, C. J. 1999. A role for the leucine-responsive regulatory protein and integration host factor in the regulation

of the Salmonella plasmid virulence (spv) locus in *Salmonella typhimurium*. Mol. Microbiol. **34**:134–145.

70. Matin, A., Auger, E. A., Blum, P. H., and Schultz, J. E. 1989. Genetic basis of starvation survival in nondifferentiating bacteria. Annu. Rev. Microbiol. **43**:293–316.

71. Maw, J., and Meynell, G. G. 1968. The true division and death rates of *Salmonella typhimurium* in the mouse spleen determined with superinfecting phage P22. Br. J. Exp. Pathol. **49**:597–613.

72. McDougald, D., Rice, S. A., Weichart, D., and Kjelleberg, S. 1998. Nonculturability: Adaptation or debilitation? *FEMS Microbiol. Ecol.* **25**:1–9.

73. McFeters, G. A., Yu, F. P. P., Pyle, B. H., and Stewart, P. S. 1995. Physiological assessment of bacteria using fluorochromes. J. Microbiol. Meth. **21**:1–13.

74. McKay, A. L., Peters, A. C., and Wimpenny, J. W. T. 1997. Determining specific growth rates in different regions of *Salmonella typhimurium* colonies Lett. Appl. Microbiol. **24**:74–76.

75. McLean, R. J. C., Whiteley, M., Hoskins, B. C., Majors, P. D., and Sharma, M. M. 1999. Laboratory techniques for studying biofilm growth, physiology, and gene expression in flowing systems and porous media. Meth. Enzymol. **310**:248–264.

76. Messer, W., and Weigel, C. 1996. Initiation of chromosome replication. In Escherichia coli *and* Salmonella: *Cellular and Molecular Biology*, pp. 1579–1601 (F. C. Neidhardt, Ed.). ASM Press, Washington, D.C.

77. Millar, M. R., Linton, C. J., Cade, A., Glancy, D., Hall, M., and Jalal, H. 1996. Application of 16S rRNA gene PCR to study bowel flora of preterm infants with and without necrotizing enterocolitis. J. Clin. Microbiol. **34**:2506–2510.

78. Moat, A. G., and Foster, J. W. 1995. Microbial physiology, 3rd ed. J. Wiley, New York. 580p.

79. Moir, A., and Smith, D. A. 1990. The genetics of bacterial spore germination. Annu. Rev. Microbiol. **44**:531–553.

80. Mukamolova, G. V., Kaprelyants, A. S., Young, D. I., Young, M., and Kell, D. B. 1998. A bacterial cytokine. Proc. Natl. Acad. Sci. USA **95**:8916–8921.

81. Mukamolova, G. V., Yanopolskaya, N. D., Kell, D. B., and Kaprelyants, A. S. 1998. On resuscitation from the dormant state of *Micrococcus luteus*. Antonie Van Leeuwenhoek **73**:237–243.

82. Neidhardt, F. C., Ingraham, J. L., and Schaechter, M. 1990. *Physiology of the Bacterial Cell: A Molecular Approach*. Sinauer Associates, Sunderland, MA.

83. Neidhardt, F. C., and Curtiss, R. 1996. *Escherichia coli* and *Salmonella:* cellular and molecular biology, 2nd ed. ASM Press, Washington, D.C. 2822p.

84. Nilsson, L., Oliver, J. D., and Kjelleberg, S. 1991. Resuscitation of *Vibrio vulnificus* from the viable but nonculturable state. J. Bacteriol. **173**:5054–5059.

85. Nystrom, T. 1998. To be or not to be: The ultimate decision of the growth-arrested bacterial cell. FEMS Microbiol. Rev. **21**:283–290.

86. Nystrom, T. 1999. Starvation, cessation of growth and bacterial aging. Curr. Opin. Microbiol. **2**:214–219.

87. Nystrom, T. 2001. Not quite dead enough: On bacterial life, culturability, senescence, and death. Arch. Microbiol. **176**(3):159–164.

88. Nystrom, T., and Neidhardt, F. C. 1994. Expression and role of the universal stress protein, uspa, of *Escherichia coli* during growth arrest. Mol. Microbiol. **11**:537–544.

89. Oliver, J. D. 1993. Formation of viable but nonculturable cells. In *Starvation in Bacteria*, pp. 239–272 (S. Kjelleberg, Ed.). Plenum, New York.

90. Oliver, J. D. 1995. The viable but non-culturable state in the human pathogen *Vibrio vulnificus*. FEMS Microbiol. Lett. **133**:203–208.

91. Oliver, J. D., and Bockian, R. 1995. In vivo resuscitation, and virulence towards mice, of viable but nonculturable cells of *Vibrio vulnificus*. Appl. Environ. Microbiol. **61**:2620–2623.

92. Opperman, T., Murli, S., Smith, B. T., and Walker, G. C. 1999. A model for a umuDC-dependent prokaryotic DNA damage checkpoint. Proc. Natl. Acad. Sci. USA **96**(16):9218–9223.

93. Pirt, S. J. 1975. *Principles of Microbe and Cell Cultivation*. Blackwell Scientific, Oxford.

94. Polissi, A., Pontiggia, A., Feger, G., Altieri, M., Mottl, H., Ferrari, L., and Simon, D. 1998. Large-scale identification of virulence genes from *Streptococcus pneumoniae*. Infect. Immun. **66**(12):5620–5629.

95. Postgate, J. 1967. Viability measurements and the survival of microbes under minimum stress. Adv. Microb. Physiol. **1**:1–21.

96. Postgate, J. R. 1969. Viable counts and viability. Meth. Microbiol. **1**:611–628.

97. Rahman, I., Shahamat, M., Chowdhury, M. A., and Colwell, R. R. 1996. Potential virulence of viable but nonculturable Shigella dysenteriae type 1. Appl. Environ. Microbiol. **62**:115–120.

98. Raivio, T. L., and Silhavy, T. J. 1999. The sigma(E) and Cpx regulatory pathways: Overlapping but distinct envelope stress responses. Curr. Opin. Microbiol. **2**:159–165.

99. Relman, D. A. 1999. The search for unrecognized pathogens. Science **284**:1308–1310.

100. Relman, D. A., Loutit, J. S., Schmidt, T. M., Falkow, S., and Tompkins, L. S. 1990. The agent of bacillary angiomatosis – an approach to the identification of uncultured pathogens. N. Engl. J. Med. **323**:1573–1580.

101. Relman, D. A., Schmidt, T. M., Macdermott, R. P., and Falkow, S. 1992. Identification of the uncultured bacillus of Whipple's disease. N. Engl. J. Med. **327**:293–301.

102. Rigsbee, W., Simpson, L. M., and Oliver, J. D. 1997. Detection of the viable but nonculturable state in *Escherichia coli* O157:H7. J. Food Safety **16**:255–262.

103. Roszak, D. B., and Colwell, R. R. 1987. Survival strategies of bacteria in the natural environment. Microbiol. Rev. **51**:365–379.

104. Russell, J. B., and Cook, G. M. 1995. Energetics of bacterial-growth-balance of anabolic and catabolic reactions. Microbiol. Rev. **59**:48–62.

105. Saier, M. H. 1998. Multiple mechanisms controlling carbon metabolism in bacteria. Biotech. Bioeng. **58**:170–174.

106. Salgado, H., SantosZavaleta, A., GamaCastro, S., MillanZarate, D., Blattner, F. R., and ColladoVides, J. 2000. RegulonDB (version 3.0): transcriptional regulation and operon organization in *Escherichia coli* K-12. Nucleic Acids Research **28**(1):65–67.

107. Sawers, G. 1999. The aerobic/anaerobic interface. Curr. Opin. Microbiol. **2**:181–187.

108. Schoedon, G., Goldenberger, D., Forrer, R., Gunz, A., Dutly, F., Hochli, M., Altwegg, M., and Schaffner, A. 1997. Deactivation of macrophages with interleukin-4 is the key to the isolation of *Tropheryma whippelii*. J. Infect. Dis. **176**:672–677.

109. Severinov, K. 2000. RNA polymerase structure-function: Insights into points of transcriptional regulation. Curr. Opin. Microbiol. **3**:118–125.

110. Sheridan, G. E., Masters, C. I., Shallcross, J. A., and MacKey, B. M. 1998. Detection of mRNA by reverse transcription-PCR as an indicator of viability in *Escherichia coli* cells. Appl. Environ. Microbiol. **64**:1313–1318.

111. Singh, A., Yeager, R., and McFeters, G. A. 1986. Assessment of in vivo revival, growth, and pathogenicity of *Escherichia coli* strains after copper- and chlorine-induced injury. Appl. Environ. Microbiol. **52**:832–837.

112. Smith, H. 2000. Questions about the behaviour of bacterial pathogens *in vivo*. Philos. Trans. R. Soc. Lond. B Biol. Sci. **355**:551–564.

113. Spector, M. P. 1998. The starvation-stress response (SSR) of *Salmonella*. Adv. Microb. Physiol. **40**:233–279.

114. Staley, J. T., and Konopka, A. 1985. Measurement of in situ activities of non-photosynthetic microorganisms in aquatic and terrestrial habitats. Annu. Rev. Microbiol. **39**:321–346.

115. Storz, G., and Imlay, J. A. 1999. Oxidative stress. Curr. Opin. Microbiol. **2**:188–194.

116. Stragier, P., and Losick, R. 1996. Molecular genetics of sporulation in *Bacillus subtilis*. Annu. Rev. Genet. **30**:297–341.

117. Suau, A., Bonnet, R., Sutren, M., Godon, J. J., Gibson, G. R., Collins, M. D., and Dore, J. 1999. Direct analysis of genes encoding 16S rRNA from complex

communities reveals many novel molecular species within the human gut. Appl. Environ. Microbiol. **65**:4799–4807.

118. Sun, Q., and Margolin, W. 1998. FtsZ dynamics during the division cycle of live *Escherichia coli* cells. J. Bacteriol. **180**:2050–2056.

119. Tao, H., Bausch, C., Richmond, C., Blattner, F. R., and Conway, T. 1999. Functional genomics: Expression analysis of *Escherichia coli* growing on minimal and rich media. J. Bacteriol. **181**:6425–6440.

120. Walker, G. C. 1996. The SOS response of *Escherichia coli*. In Escherichia coli *and* Salmonella: *Cellular and Molecular Biology*, pp. 1400–1416 (F. C. Neidhardt, Ed.). ASM Press, Washington, D.C.

121. Wassarman, K. M., Zhang, A. X., and Storz, G. 1999. Small RNAs in *Escherichia coli*. Trends Microbiol. **7**:37–45.

122. Whitesides, M. D., and Oliver, J. D. 1997. Resuscitation of *Vibrio vulnificus* from the viable but nonculturable state. Appl. Environ. Microbiol. **63**:1002–1005.

123. Wilson, K. H., Blitchington, R., Frothingham, R., and Wilson, J. A. P. 1991. Phylogeny of the Whipple's-disease-associated bacterium. Lancet **338**:474–475.

124. Wilson, K. H., and Blitchington, R. B. 1996. Human colonic biota studied by ribosomal DNA sequence analysis. Appl. Environ. Microbiol. **62**:2273–2278.

125. Wimpeny, J. W. T. 1988. The bacterial colony. In *CRC Handbook of Laboratory Modelling Systems for Microbial Ecosystems*, pp. 109–139. CRC Press, Boca Raton, FL.

126. Xu, H. S., Roberts, N., Singleton, F. L., Attwell, R. W., Grimes, D. J., and Colwell, R. R. 1982. Survival and viability of nonculturable *Escherichia coli* and *Vibrio cholerae* in the estuarine and marine environment. Microb. Ecol. **8**:313–323.

127. Yura, T., and Nakahigashi, K. 1999. Regulation of the heat-shock response. Curr. Opin. Microbiol. **2**:153–158.

128. Zambrano, M. M., and Kolter, R. 1996. GASPing for life in stationary phase. Cell. **86**:181–184.

129. Zhou, P., Bogan, J. A., Welch, K., Pickett, S. R., Wang, H. J., Zaritsky, A., and Helmstetter, C. E. 1997. Gene transcription and chromosome replication in *Escherichia coli*. J. Bacteriol. **179**: 163–169.

130. Zinser, E. R., and Kolter, R. 2000. Prolonged stationary-phase incubation selects for lrp mutations in *Escherichia coli* K-12. J. Bacteriol. **182**(15):4361–4365.

CHAPTER 2

Survival of environmental and host-associated stress

Petra Dersch and Regine Hengge-Aronis

SPECIFIC AND GENERAL STRESS RESPONSES IN BACTERIA

In their unicellular state, bacteria are directly exposed to frequent and sometimes dramatic changes in an environment. The only way to survive and, if possible, to grow and multiply, is to adapt to these changes. Thus, being exposed to stressful conditions is the normal way of life for bacteria. Sometimes adaptation involves global changes in gene expression, which have made stress responses the paradigm of global gene regulation (130).

The necessity to adapt to environmental stresses applies to non-pathogenic and pathogenic bacteria alike. However, for a pathogen, stress responses are also crucial for coping with specific host-inflicted stress situations once the pathogen has entered the host organism. From the perspective of the pathogen, various host tissues or cellular compartments are characterized by different stress conditions. Therefore, a pathogen can also use its ability to sense environmental stresses as a guidance system and as a device to prepare in good time for upcoming stresses in its journey through the host organism.

In principle, two types of stress response have to be distinguished in bacteria: stress-specific responses and the general stress response. The numerous specific responses are induced by defined single-stress conditions, such as certain reactive oxygen species or specific DNA damage. The physiological function of these responses is to cope with the actual stress situation only. This may include elimination of the stress agent as well as repair of damage that has already occurred. By contrast, a general stress response is in principle a single far-reaching response that can be triggered by many very different stress signals (62, 109). Its main physiological function is not so much the repair of damage which has already occurred, but rather to render the cells broadly stress-resistant, even against stresses not yet experienced (cross-protection).

Thus, the general stress response is mainly a preventive response and may reflect the fact that, in contrast to the artificial laboratory world, one stress rarely occurs alone in nature. One of the most interesting features of general stress responses is their enormous signal integration potential.

A major signal for the induction of the general stress response is the scarcity or even absence of nutrients (65, 78). Therefore, induction of the general stress response is often associated with a reduction in growth rate, or even with a cessation of growth. The latter is accompanied by a transition to a state of reduced metabolism directed towards maintenance rather than rapid growth. In the extreme, this may lead to a (semi)-dormant state. Thus, the stationary phase response is part of the general stress response. Nevertheless, this is just one aspect of the general stress response, which can also be induced by other signals (e.g., increased osmolarity or high temperature), which do not necessarily interfere with growth.

Strikingly, a list of stress signals that induce the general stress response in the gram-negative bacterium *Escherichia coli* (for which stress responses are best studied) reads like a list of conditions prevailing in different parts of a mammalian host organism. These include scarcity of various nutrients, acidic pH, high osmolarity, classical heat shock conditions (42° C), and high cell density of competing bacteria. In addition, stress agents such as reactive oxygen species that do not induce the general stress response (at least in *E. coli*) are encountered by the pathogen within the host organism. Taken together, this indicates that the general stress response as well as specific stress responses should be induced in pathogens when they invade and colonize a host organism. In addition, the stationary phase response in particular, which may result in a kind of dormant state, could be important for pathogenic bacteria in nutrient-poor intra-host environments or in chronic infections. Moreover, there is now clear evidence that bacteria use their stress responses, i.e., their ability to process information about their immediate environment within a host, in order to regulate their virulence genes appropriately (70). In other words, specific and general stress responses and the control of virulence genes in pathogenic bacteria are tightly interconnected.

STATIONARY PHASE AND GENERAL STRESS SURVIVAL STRATEGIES

Readily Reversible Stationary Phase Responses

The absence of essential nutrients and therefore entry into stationary phase can trigger either a long-term genetic program, which may lead to differentiation, i.e., sporulation, or a more flexible and rapidly reversible response. Which type of response occurs is probably a function of the probability of nutrients becoming available soon in the specific habitat.

Enteric bacteria such as *E. coli* are typical examples of non-sporulating bacteria. During entry into stationary phase, *E. coli* cells induce the general stress response which results in multiple stress resistance, morphological alterations that generate smaller ovoid cells, the production of storage compounds (glycogen, polyphosphate) and numerous metabolic changes (65, 66, 78). Underlying this response is a complex hierarchical network with the σ^S (RpoS) subunit of RNA polymerase (RNAP) at the top (62, 65, 66, 83, 88). Even though this response involves a complex genetic program, it is completely and rapidly reversible at any time. This may reflect the fact that enteric bacteria lead a "feast or famine" existence characterized by sporadic availability of nutrients, for which there is intense competition. This response (which, for instance, is equally induced by low temperature) also promotes survival in the environment outside the host, where nutrients can be very scarce, and in this situation oxidative stress resistance generated by the general stress response can be especially crucial for survival.

Some sporulating bacteria, such as *Bacillus subtilis* can choose between inducing a flexibly reversible general stress response or entering the sporulation program when exposed to starvation for an essential nutrient. The *B. subtilis* general stress response is dependent on the σ^B subunit of RNAP (109). σ^B belongs to the alternative sigma factors of the ECF family. σ^B is activated and, due to positive autoregulation, is also induced in response to almost the same series of stress conditions as σ^S in enteric bacteria. However, the mechanism of σ^B regulation is completely different than that of σ^S and involves an anti-sigma factor and an anti-anti-sigma factor. The activity of the latter is dependent on its state of phosphorylation, which in turn is controlled by several converging stress signal pathways (109). The target genes activated by σ^B-containing RNAP include a number of genes whose homologs in *E. coli* are induced by σ^S-containing RNAP (53). Thus, the general stress responses in these gram-positive and gram-negative model organisms are triggered by the same signals and produce the same overall physiological output. However, the molecular mechanisms of signal transduction and integration and the regulatory proteins that link stress signal input and physiological output are different.

Recent data indicate that the σ^B-dependent general stress response is important for virulence of gram-positive pathogens such as *Staphylococcus* species (22, 79; see below). The basic components of this regulation by σ^B are very similar to those in *B. subtilis* but the signal input pathways seem to be different, which is in agreement with the two species living in very different environments, i.e., one associated with a mammalian host and the other in the soil, respectively.

Sporulation

Sporulation is a developmental program that is usually triggered by nutrient starvation in combination with high cell density. Sporulation results in the formation of a dormant cell form, i.e., a spore, that is morphologically different from a vegetatively growing cell and is usually characterized by extreme stress resistance and longevity (124, 128). Unlike the induction of the readily reversible general stress response, the sporulation program involves a commitment, i.e., once beyond a certain point in the program the cell has to complete the full sporulation and germination cycle when nutrients becoming available again. This explains why sporulation occurs in, for example, soil environments where starvation tends to be a long-term phenomenon. This does not mean that pathogenic bacteria are, in principle, non-sporulating. Pathogens such as *Bacillus anthracis*, which spend part of their life cycle in an outside environment where they can be exposed to long-term starvation and other severe environmental assaults, are spore formers. However, within the host this seldom occurs, because of the relative abundance of nutrients.

High Cell Density Responses: Quorum Sensing

Certain activities only make sense in the microbial world when they are performed by many individual organisms simultaneously. The now classic example of such an activity is the production of light by the *lux* system of marine bacteria (46, 47, 58). The same, however, applies to the expression of virulence factors that can only overwhelm host defense mechanisms when produced in sufficient quantities (30). This means that bacteria can sense their own population density. They do so by a process called "quorum sensing," which involves the excretion of small molecules called autoinducers (often acylated homoserine lactones in gram-negative species, and small peptides in gram-positive bacteria) into the medium during growth (46, 139). Once accumulated beyond a certain threshold concentration, which reflects a certain cell density in the culture, the signalling molecule can trigger the expression of various quorum sensing–controlled genes by directly or indirectly affecting the activity of a key regulatory protein (e.g., LuxR in the *V. fischeri* system). Since in a laboratory set-up, high cell densities often coincide with entry into stationary phase, induction of gene expression under these conditions requires careful analysis of the inducing signal, which may be starvation for an essential nutrient, reduction of the growth rate or quorum signaling. The details of quorum sensing and its role in the control of pathogenicity are covered in detail elsewhere in this volume (see Chapter 3).

THE σ^S-DEPENDENT STATIONARY-PHASE AND GENERAL STRESS RESPONSE

Physiology

The general stress response, which includes the stationary phase response, crucially contributes to the global control of virulence gene expression. The general stress responses of *E. coli* and *B. subtilis* have been studied in great detail. In *E. coli* and related bacteria (i.e., the γ-branch of proteobacteria) the connections to pathogenicity are especially obvious. Therefore, the *E. coli* general stress response is introduced here as a model system. Most of what we know about the physiology of the general stress response is based upon the analysis of stationary phase cells. The general stress response is flexible enough to serve both as a rapid and sometimes transient emergency response (42) and as a long-term program of adaptation to starvation and other stresses, which can involve dramatic changes in cellular physiology and morphology (62, 65, 66, 88).

The general stress response is induced when *E. coli* cells are starved of an essential source of carbon, nitrogen, phosphorus or amino acids. Moreover, shifts to high osmolarity, low or high temperature (20° C and 42° C, respectively) or low pH are efficient inducing conditions. Under all these conditions, the cellular concentration of the master regulator σ^S (RpoS), a promoter-recognizing subunit of RNA polymerase, significantly increases (51, 81, 87). The general stress response is usually (but not always) accompanied by a reduction or cessation of growth and provides the cell with the ability to survive the actual stress as well as additional stresses not yet experienced (cross-protection). This is best described for stationary phase cells, which become strongly resistant against high temperature (>48° C), various types of oxidative stress, high salt concentrations (e.g., 3 M NaCl), acidic pH or near-UV irradiation (73, 74, 83, 95). Stress-induced proteins that contribute to this multiple stress resistance include several catalases, exonuclease III (a DNA repair enzyme), glutathione reductase, a DNA-protecting histone-like protein (Dps), a periplasmic superoxide dismutase, the trehalose-synthesizing enzymes, two glutamate decarboxylases and others. The expression of all these factors requires the master regulator σ^S (summarized in (62)).

Moreover, stationary phase cells or cells which have been subjected to high osmolarity change their morphology; for example they become smaller and ovoid-shaped rather than rod-shaped (82, 120). This phenotype is controlled by the BolA protein, which in turn is expressed from a σ^S-dependent gene (2, 82). Since many σ^S-dependent stress-inducible genes code for extra-cytoplasmic proteins, the composition and/or other properties of the membranes and the periplasm are altered upon induction of the general stress response (13, 55). The same applies to certain metabolic pathways, such as pyruvate/acetate metabolism (23). Finally, the expression of a number of fimbrial genes (3, 118, 119) and other virulence genes is typically altered in stationary phase or under other stress conditions that result in increased cellular σ^S levels (see below for more details).

The overall consequence of these alterations is a cell which is well adapted to cope with whatever threatening conditions it may encounter in the outside environment or within a host organism. For the time being this cell concentrates on survival under all circumstances, although it can return to rapid growth and multiplication at any moment if conditions improve.

Regulation of σ^S by Environmental Signals

The basis of the general stress response in *E. coli* is the rapid increase in the cellular level of the master regulator σ^S under quite different stress conditions. Thus, a major question with respect to σ^S regulation concerns the integration of many different signals towards a single parameter, the cellular σ^S concentration. This integration is based on different signals affecting different levels in the control of σ^S, for example the transcription of the *rpoS* gene, the translation of *rpoS* mRNA and the rate of proteolysis of the σ^S protein (81). Certain stress conditions, which affect more than one of these processes simultaneously, such as high osmolarity or low pH, which both stimulate *rpoS* translation and inhibit σ^S degradation, have an especially strong and rapid inducing effect. This signal integration and regulation, which involves dozens of regulatory factors, makes σ^S the most complexly regulated *E. coli* protein that has been identified so far (64, 65a).

Of the σ^S control mechanisms, the regulation of *rpoS* transcription is the least well understood. Under non-stress conditions, when cells grow continously in minimal media, *rpoS* mRNA is synthesized constantly. *rpoS* transcription is not autoregulated (81). It is induced by a gradual reduction in growth rate, for instance during entry into stationary phase in a rich medium (81). It is also positively and negatively controlled by the alarmone guanosine-tetra-phosphate (ppGpp) and the cyclic AMP–CRP complex, respectively (80, 81). The molecular details of this regulation have not been studied.

The basis of translational regulation of *rpoS* is most likely to be related to its mRNA secondary structure. The translational initiation site (including the ribosomal binding site and the start codon) are probably base-paired and therefore not easily accessible to ribosomes (17, 100). It is hypothesized that this secondary structure is melted in response to certain stress signals and then allows access of ribosomes (84, 91), although this has not yet been demonstrated directly. The regulation of translation is obviously a tightly controlled process involving many different regulatory factors. These include the proteins Hfq and HU, which both bind to rpoS mRNA in vitro (6, 100), and the protein H-NS, which has a strongly inhibiting effect on *rpoS* translation (7, 140). Interestingly, both HU and H-NS are histone-like proteins better known for their association with the *E. coli* nucleoid DNA. Furthermore, at least three small regulatory RNAs are involved in the control of *rpoS* translation. These include DsrA, which is partially complementary to the region of *rpoS* mRNA, which intramolecularly base-pairs to the translational initiation region and is itself transcriptionally induced during growth at low temperature (84, 91, 111, 127). This suggests, at least theoretically, how low temperature can result in translational induction of *rpoS*. A second small RNA that can positively influence *rpoS* translation is RprA (90). However, both its physiological role and its mechanism of action remain obscure, since RprA affects *rpoS* expression only when overproduced and exhibits weak sequence complementarity to *rpoS* mRNA only. The OxyS RNA is specifically induced under hydrogen peroxide stress and inhibits *rpoS* translation, probably by forming a ternary complex with *rpoS* mRNA and Hfq protein, which is translationally unproductive (142). Finally, factors that indirectly affect *rpoS* translation, e.g., the regulatory protein LeuO which inhibits dsrA transcription (76), are also involved. All this evidence suggests that a complex network involving numerous signal transducing and regulatory factors controls *rpoS* translation in response to various stress signals. However, a coherent picture of how all these factors work together has yet to emerge.

Equally important is the regulation of σ^S proteolysis. During growth on a minimal medium, a highly dynamic equilibrium between a certain rate of σ^S synthesis and σ^S degradation can be observed. The result is a low but measurable σ^S level in the cell, which can rapidly increase either by enhanced translation (see above) or by a sudden reduction in σ^S proteolysis. The advantage of this regulation is its extremely rapid kinetics unrivalled by any other type of regulation. The degradation of σ^S is dependent on the ClpXP protease (123) and the recognition factor RssB, which belongs to the two-component response regulator family of proteins (8, 99, 108). Only the phosphorylated form of RssB exhibits high affinity for σ^S. A small region in σ^S, region 2.5, which probably folds into an amphiphilic alpha helix, is sufficient for efficient binding to phosphorylated RssB. A single amino acid in the 2.5 region (lysine 173) is absolutely crucial for this recognition (10). The entire RssB protein (which consists of at least two domains) is required for σ^S binding, perhaps because both domains may contribute to a binding pocket for region 2.5 in σ^S (77). The RssB/ClpXP system seems to be a relatively simple recognition and degradation machinery but nevertheless is able to integrate numerous signals that control the rate of σ^S turnover (64). The molecular details of the stress signal pathways linked to the RssB/ClpXP system have yet to be elucidated, but there are many possibilities for stress signal input. First, as a response regulator, RssB phosphorylation and dephosphorylation is likely to be controlled by certain stress signals. Second, RssB is the limiting factor for σ^S proteolysis in vivo. Therefore, σ^S stability can in principle be increased by either downregulating RssB or upregulating the synthesis of σ^S (110a). The latter should result in titration of the recognition system and actually occurs in response to shifts to high osmolarity or acidic pH. Third, binding of σ^S to RNAP core enzyme protects σ^S against degradation (143). Therefore, any factor that stimulates this association should not only activate but also stabilize σ^S. Finally, σ^S could also be stabilized under conditions where recognition by RssB is not affected but where ClpXP may be inactivated or "monopolized" by increasing cellular concentrations of substrates competing with σ^S for ClpXP (64).

In summary, the enormous signal integration potential of the complex σ^S regulatory system is the basis for the far-reaching physiological functions of σ^S under nearly all adverse environmental conditions.

Regulation of Gene Expression by σ^S

More than 70 genes or operons have been identified as being under direct or indirect control of σ^S (see (89) for a recent compilation). In many cases, additional regulatory proteins participate in the regulation of these genes. In particular, histone-like proteins such as H-NS, Fis, integration host factor (IHF) or leucine-responsive regulatory protein (Lrp) are often involved, but also more specifically acting regulatory factors such as cAMP–CRP can play a role (63). The result is a complex regulatory network with σ^S as the essential master regulator at the top, which exhibits intricate fine regulation of the many member genes. This network also includes regulatory cascades established by secondary regulators, which themselves are under σ^S control and can integrate additional signal input. Thus, smaller groups of genes within the σ^S network are grouped together in regulatory modules that are not only controlled by the signals that regulate σ^S but also are characterized by additional module-specific signal processing.

The textbook concept of prokaryotic gene regulation by various sigma factors is that alternative sigma factors are activated or induced under certain conditions, replace the vegetative sigma factor and thereby reprogram RNAP to activate other promoters. This concept is based on the idea that each sigma factor recognizes a specific and distinct promoter sequence. However, σ^S does not seem to follow this principle. While σ^S and the vegetative sigma factor σ^{70} activate very different genes in vivo (62), they recognize basically the same promoter sequences in standard in vitro transcription assays (101, 133, 134). So what is the mechanism of σ^S promoter selectivity in vivo? The core promoter consensus sequences, namely the −35 and −10 regions that can be derived by comparison of in vivo σ^S-controlled promoter sequences and that yield maximal expression in vitro, are indeed identical for RNAP containing either σ^S or σ^{70}. Nevertheless, there are a number of minor differences, which in combination can generate σ^S-selectivity: (i) in natural σ^S-dependent promotors, the -35 region is often strongly degenerate but can still be used by σ^S-containing RNAP (9, 36); (ii) a cytosine at position −13 (located just upstream of the core −10 region) strongly and selectively stimulates σ^S-mediated expression, because it is contacted by a specific amino acid in σ^S (K173) not present in σ^{70} (9); (iii) the region downstream of the −10 region is AT-rich in almost all known σ^S-dependent promoters (9), and this feature contributes to σ^S-dependent induction in stationary phase (110a). In addition to these sequence-based signals in a promoter region that are likely to be decoded by σ^S-containing RNAP itself, additional regulator proteins, which bind to specific sequences flanking the promoter region, can contribute to σ^S-selectivity. Either such regulators can activate more strongly with σ^S-containing RNAP than with RNAP complexed with σ^{70} or they repress more strongly in conjunction with σ^{70}-containing RNAP (26, 52).

In conclusion, the analysis of σ^S promoter selectivity has modified our general concept of prokaryotic promoter recognition and activation in a way that now includes the formation of sometimes large and always promoter-specific nucleoprotein complexes. These structures certainly contribute to the complexity and flexibility of gene regulation under various stress conditions.

SURVIVAL STRATEGIES UNDER HOST-INDUCED STRESS CONDITIONS

Counteracting Specific Host-Associated Stresses

Host organisms are the evolutionary products of a long-term interrelationship with microbes that surround them. In order to compete and to gain advantage in this coexistence, hosts have evolved a complex set of overlapping defense systems. Pathogenic microbes have to first pass physical (mucus) and chemical (secreted enzymes, complement systems) barriers to colonize and enter host tissues. Upon entry into host tissues they have to face attack by the immune system directed to eliminate invaders (oxidative burst) or must cope with other deleterious environmental conditions such as acidic pH and limiting amounts of essential nutrients such as iron. A successful pathogen thus relies on its ability to sense host-induced stress conditions and to respond with specific counteracting strategies.

Oxidative Stress Response

Among the most sophisticated bactericidal defense mechanisms of higher organisms are components of the unspecific immune system: professional phagocytes, such as macrophages and polymorphic neutrophils (PMNs). They detect foreign microbes, enclose them and ingest them in a membrane-enclosed compartment (phagosome) in which they are destroyed and eliminated. The deleterious effect of phagocytes is mostly based on the production of reactive oxidative intermediates (oxidative burst, O_2-dependent defense mechanisms) and hypochloride (OCl⁻) that damage and destroy microbial lipids, proteins and nucleic acids.

To prevent killing by phagocytes, pathogens evolved effective strategies to sense and inactivate oxygen radicals. A large number of newly synthesized proteins have been detected in bacterial pathogens upon oxidative stress in vitro and in a macrophage environment in vivo. These proteins can be divided into those protecting against peroxide stress (H_2O_2), e.g., catalases, and those which protect against superoxide stress (O_2^-), e.g., superoxide dismutases (SODs) (41). Both enzyme classes have antioxidant properties and either eliminate or convert oxidative substances. The synthesis of a subset of these enzymes is induced by the transcriptional activator protein OxyR; another set is under the control of the activator proteins SoxS and SoxR. OxyR and SoxR are both sensor and regulator proteins of the H_2O_2 response, which undergo conformational changes when oxidized by H_2O_2 or O_2^- (131). This alteration changes their specific contacts with DNA and leads to a rapid activation of oxidative stress-induced promoters (33). The fact that many genes of the oxidative stress response are induced in macrophages raised the question of whether the oxidative stress response is important for virulence. In fact, *E. coli soxRS* mutants are hypersusceptible to killing by murine macrophages (103), and SOD defective strains of *Y. enterocolitica, Helicobacter pylori* and the Gram-positive organism *Streptococcus pneumoniae* are hypersensitive to oxidative stress and are strongly attenuated in virulence (116, 125).

Another important factor that enables bacterial pathogens to survive the oxidative burst in host tissues is their general capacity to efficiently repair oxidative damage of DNA and proteins. Besides special H_2O_2-induced exonucleases, bacterial pathogens use the RecA recombinase as a major repair protein of oxidative DNA damage. The necessity of this repair system for pathogenicity became especially apparent with different *recA* mutant strains of *Salmonella* and *E. coli* EHEC, which exhibit a dramatically reduced virulence in mice (19, 45).

In addition to these responses specifically induced by reactive oxygen species, induction of the general stress response e.g., by starvation or hyperosmolarity, dramatically increases oxidative stress resistance (74, 83). This may be important to a pathogen, which is suddenly exposed to life-threatening oxidative bursts produced by macrophages. In fact, several structural genes, which contribute to oxidative stress resistance (e.g., katG and dps, which encode a catalase and a DNA-protecting protein, respectively), can be activated either by oxidative stress (via the OxyR system) or by the general stress response (via σ^S). Thus the specific oxidative stress response is tightly interwoven with the stationary phase response.

Acidic Stress Response

Another common environmental threat endured by pathogens is acidic stress. During their existence and travel within a host, microorganisms encounter very low pH in the stomach, fatty acids in the gut and urea in the urogenital tract. Furthermore, they must endure the acidic environment in the macrophage and PMN phagosomes. Enteric bacteria possess acidic stress response mechanisms that protect the bacteria at pH 3 for several hours (43). Acidification by organic acids or by lowering the pH results in the induction of more than 50 acid shock proteins that help to adapt and keep the internal pH relatively constant. In Gram-negative bacteria, this is usually achieved by the activation of proton pumps, such as K^+/H^+ or Na^+/H^+ antiporters (12). Alternatively, Gram-positive bacteria increase the level of F_0F_1 proton-translocating ATPase to adjust their internal pH. Another acid resistance mechanism requires the enzyme glutamate decarboxylase that converts intracellular glutamate to γ-amino butyric acid, thereby consuming protons, and the glutamate/GABA antiporter, expelling the decarboxylation product (43). Several regulatory proteins have been identified that play a role in the acidic stress response in pathogens. One of these is the alternative sigma factor σ^S (see above), which is absolutely required to survive acid stress imposed by volatile fatty acids (i.e., the condition in some parts of the intestine) and significantly contributes to inorganic acid tolerance (i.e., conditions in the stomach). Other regulators that control subsets of acid stress proteins are the iron regulator Fur and the PhoPQ system ((15, 54); see also below).

The impact of acidic stress on virulence of enteric bacteria is profound. A significant number of virulence genes are known to be induced in response to the low pH environment of the phagolysosome, where the pH drops to about 4–5 within 60 minutes after formation. Several mutant strains of *Salmonella typhimurium*, *Shigella flexneri* and *Listeria monocytogenes* defective in acid tolerance are also significantly attenuated in virulence (50, 92, 112). Hence, it has been suggested that the acid stress response contributes to the pathogenesis of an organism. This hypothesis is supported by the finding that clinical isolates of hemorrhagic *E. coli* (EHEC O157:H7) are extremely acid resistant (18). In the case of *H. pylori*, a pathogen colonizing the human stomach, acid resistance is a prerequisite for its survival (93, 106). One major mechanism of this pathogen, to adapt to low pH, involves the enzyme urease that converts urea to CO_2 and ammonia and neutralizes low pH. *Helicobacter* also produces special acid-induced gene products (i.e., CagA and WcbJ) which contribute to the acid tolerance by affecting the conformation and function of cell surface proteins.

Iron Deprivation

Microbes living within humans or animals face the problem of how to acquire sufficient amounts of iron to conserve the function of essential proteins, such as hemes and cytochromes. Their hosts solubilize iron mainly by binding it to highly efficient iron-sequestering molecules such as ferritin, transferrin, lactoferrin and hemoglobin in secretory fluids or intracellular storages. This reduces the amount of free iron available for bacteria to less than 1 ion per liter. Bacterial pathogens bridge the gap between demand and supply by the synthesis of multiple high-affinity iron chelators (siderophores) and corresponding high-affinity siderophore uptake systems (31). Alternatively, some pathogens can also utilize iron-loaded proteins of their hosts (e.g., hemoglobin) and bind them to specific receptors that release the iron into the bacterial cell (16, 57).

In Gram-negative organisms, expression of iron chelators and their uptake systems and the expression of virulence factors are often coregulated by the common iron regulatory protein Fur (*f*erric *u*ptake *r*egulation) in response to iron availability. This connection between iron supply and bacterial virulence has been demonstrated for a number of pathogens and seems to be important for bacterial pathogenicity. For instance, *Yersinia enterocolitica* and *Yersinia pestis* only kill mice when they synthesize yersiniabactin and a related iron-repressible outer membrane transport protein (60). Similarly, in *Neisseria gonorrhoeae*, *Legionella pneumophila*, *S. typhimurium* and *Pseudomonas aeruginosa*, specific iron assimilation systems are known to be essential for intracellular infection and pathogenicity (16). Several Fur-like regulatory proteins have been identified in pathogens and their participation in the derepression of toxin genes in response to iron deficiency has been demonstrated. Prime among these toxins are exotoxin A of *P. aeruginosa* and the diphtheria toxin from *C. diphtheria* (14, 110). Presumably, toxin synthesis that results in the destruction of eukaryotic cells might offer a new source of iron for the pathogens.

Taken as a whole, intensive research on host-associated stress responses reveals a large number of specific stress-induced factors involved in virulence and provides insight into the diversity, complexity and interconnections of stress survival and virulence. However, the global network of all stress-regulated and stress-dependent mechanisms involved in pathogenesis is far from being understood.

General Defensive Strategies of Pathogens

Pathogens also protect themselves against environmental stresses and host defense mechanisms by a number of general defense strategies. A simple and effective strategy to subvert host defense mechanisms is by biofilm or capsule formation. These specialized cell envelope structures are usually composed of polysaccharides or protein layers and form a protective shield around the bacterium that inhibits phagocytosis by professional phagocytes and confers resistance to the complement system, antibodies and antibiotics (98, 132). The formation of these surface structures is a complex process that is regulated by cell density (quorum sensing; see Chapter 3 in this book).

In order to survive, pathogenic microorganisms also secrete enzymes, such as IgA-proteases or catalases, which directly degrade essential components of the host defense system or are important for the detoxification of toxic metabolites produced by granulocytes and macrophages. Alternatively, they manipulate the structure and function of phagocytes by injecting effector proteins into the cell cytosol, which interfere with the signal transduction pathways that induce the oxidative burst or phagocytosis (type III secretion systems; (11, 28)).

Another strategy to evade identification by the immune system is to present surface structures that either mimic molecular antigenic structures of the host or are highly variable antigenic surface determinants. The different variants of the surface molecules then either differ in the presence or absence of a certain molecule (phase variation) or in their antigenic properties (antigen variation) (115). These processes can also be stimulated by enhancing the frequencies of mutation or recombination of gene fragments (a more detailed description of these strategies is given in Chapter 5 of this book).

Expression of Specific Virulence Factors

The infectious cycle of pathogens starts by the colonization of host tissue. After colonization, pathogens frequently invade host cells and multiply in the cytoplasm or persist in a membrane-bound vacuole (bacterial phagosome). Subsequently, they often damage or destroy the cell by the production of cell signalling inhibitors or toxins. For all these steps during infection, pathogens produce and secrete specific virulence factors that mediate certain pathogenic activities, i.e., colonization, cell invasion etc.

Adhesins and Invasins

Colonization of host tissue usually starts with the interaction of pathogens with special host cell receptors. This interaction is mediated by specialized adhesive structures, so-called adhesins, localized at the cell surface of the microorganism. These proteinaceous structures normally interact with proteins, glycoprotein or lipoprotein receptors on epithelial or endothelial cells, as well as leukocytes, granulocytes or extracellular matrix (ECM) proteins. Two types of adhesins are generally distinguished, fimbriae and non-fimbrial adhesins.

Fimbriae (or pili) are long cell structures that are composed of several hundred fimbrial protein subunits. The major subunit forms the main shaft of the pilus, whereas the minor subunits are intercalated between the major pilus shaft protein or form a pilus tip structure. In both cases the minor subunits constitute the adhesive structure that interacts with carbohydrate residues of glycoproteins or glycolipids.

A large number of the non-fimbrial adhesins, including some pili, not only mediate cell binding but also can promote internalization into the host cell (104). Cell invasion enables bacteria to enter and spread into deeper host tissue and constitutes one of the most important strategies of pathogens to survive host-associated stress responses and defense mechanisms. To initiate cell entry, adhesins of pathogens interact with host cell receptors involved in a variety of essential host cell functions. For instance, the invasin protein of enteropathogenic *Yersinia* binds to receptors of the β_1-integrin family involved in cell–matrix interaction, cell–cell communication and cell migration (72). Interaction of the bacterial ligand with the cell receptors induces a signal transduction cascade in the host cells, leading to rearrangements of the cytoskeleton and pseudopod formation. Circumferential binding of the pseudopods around the surface of the bacterium (zipper mechanism) then internalizes the microbe into a membrane-bound phagosome (71). In contrast, *Salmonella* induces internalization by a trigger mechanism. It injects various effector proteins into the host cell thereby triggering the formation of membrane ruffles, which leads to the internalization of bacteria in the vicinity of the host cell surface (48). After internalization, pathogens can either persist in the phagosome or use specially synthesized enzymes to lyse the vacuole and enter into the cytosol. This not only enables the bacteria to use intracellular nutrients but also promotes their escape from the bacteriocidal oxidative and acidic environment of a phagolysosome.

Toxins and Modulins

Some pathogens synthesize toxins which belong to the most effective virulence factors. They can strongly damage or kill the host. The role of toxins in the pathogenesis of microbes is not fully understood. It has been speculated that the lysis of host cells, especially hemolysin action on red blood cells, might increase the amount of iron and other nutrients available for the pathogen. Furthermore, bacteria that grow in wounds and other tissues produce toxins to destroy host tissue, to enter cells or the blood stream and kill phagocytes in the vicinity of the pathogen. There are three general types of secreted toxins (exotoxins). These are (i) A–B toxins (with A being the catalytical domain and B being responsible for toxin import into a host cell), (ii) membrane-disrupting toxins that can form pores in the host cell membrane and (iii) superantigens, which "crosslink" the major histocompatibility complex (MHC) class II of antigen-presenting cells (macrophages) and the T-cell receptor interacting with MHCII molecules. As a result, T-cells proliferate and produce cytokines, which then circulate in the blood stream and cause a plethora of symptoms (61). In the past few years, an increasing number of microbial components (including adhesins and LipidA/endotoxin) have been discovered that trigger the production of various cytokines such as IL-1, IL-6, IL-8, TNFα and IF-γ and affect the host by modulating its immune system. These virulence factors, so-called modulins, play a significant role in host defense (for more details, see Chapter 5 in this book).

Type III Secretion Systems – Effector Proteins

Most bacterial effector proteins that participate in adaptation and survival in a host environment are incorporated into the cell envelope, secreted into the medium or translocated into host cells by four different types of protein secretion systems. Type I transport systems consist of two special pore-forming proteins involved in hemolysin and RTX toxin secretion, whereas type II and type IV use components of the general secretory pathway of bacteria. The most important secretion system for translocated virulence factors is the type III secretion system (68, 85). It was found in human, animal and plant pathogens and consists of about 20 proteins that are highly conserved between the different species. One well characterized type III secretion system is that of *Yersinia* spp., which injects various effector molecules into phagocytic cells, thereby preventing phagocytosis and lysis by neutrophils and macrophages (28). The type III secretion proteins are associated with the bacterial membrane, where they form a secretion pore complex with a short hair-like structure that resembles an injection needle. Structurally, these systems are related to the flagella motor apparatus including the hook. Proteins secreted and translocated by this machinery can be divided into three catagories (68): (i) the effector proteins that are directly translocated into the host cell, interfere with signaling processes and can affect cytoskeleton, phagocytosis, apoptosis, the oxidative burst and gene transcription; (ii) proteins that form a pore in the eukaryotic cytoplasmic membrane or play a role as chaperones that keep the effectors competent for the secretion process; and (iii) proteins that have regulatory functions for the export machinery. Type III systems are characterized by tight regulation of secretion and gene expression of the transport components by various environmental signals (i.e., host cell contact, temperature etc.) (28, 68). Thus, finely tuned environmental control allows the pathogen to determine its exact location in the host and enables the microbe to limit the production of energy-consuming virulence factors and transport systems to the place of action.

REGULATION OF VIRULENCE GENES BY STRESS SIGNALS

Induction of Virulence Gene Expression by Environmental Stress Conditions

Previous sections of this chapter have illustrated the array of general mechanisms and specialized strategies used by pathogenic bacteria to cope with unfavorable conditions and to survive host defenses. Since the expression of these mechanisms requires energy, it should be limited to conditions where they are needed. The necessity to compete with other microorganisms is an additional selective pressure to optimize virulence gene expression. Most pathogens alternate between an intra-host and external existence or between different hosts. Being directly exposed to the frequently hostile changes in their various environments, pathogens need to respond quickly by modulating their expression profiles. As a consequence, genes contributing to environmental adaptation as well as to pathogenicity are generally activated in response to signals of external origin (97) (Fig. 2.1). A large variety of physical (temperature, pH, osmolarity, oxygen) and chemical parameters (ions, nutrient content, toxic substances) vary between different habitats outside and inside the host. Pathogens generally use these signals to sense which microenvironment they currently occupy and to alter virulence gene expression. Thus, it is not surprising that expression of particular virulence factors is linked to various stress-associated global control circuits. In recent years, it became apparent that there are major overlaps between various specific and general environmental stress responses, as induced by oxidative stress, heat shock, starvation, high cell density, acidic stress, and iron deprivation. As a result, certain stress-responsive genes as well as virulence genes can be induced by different environmental factors and can be under the control of several stress regulons. The following section focuses on various specific and general regulatory mechanisms used by pathogens to respond to environmental signals and stress conditions and to control virulence gene expression appropriately.

Specific Regulator Proteins of Virulence Factors

Transcriptional Activators

Many virulence genes are regulated by transcriptional activator proteins whose activities are modulated by certain environmental (stress) conditions. These activators fall into different families that derive their names from a prototypic member. Especially common are regulators of the AraC and LysR types.

AraC-like activators constitute the largest known family of activators with more than 100 members (49). AraC-like regulators can be divided into functional activator subclasses, one involved in carbohydrate metabolism (cellubiose, melibiose) and the others regulating environmentally controlled stress genes (responding to heavy metals, antibiotics, oxidative stress) and virulence genes. This subset of AraC-type regulators is involved in the production of virulence factors in infections of mammals and plants and all of them respond to different environmental stimuli. AraC-like activators contain a typical helix-turn helix motif (HTH) at their conserved C-terminal domains, which binds to specific operator sequences upstream or overlapping the −35 region of gene promoters. AraC-like activators usually induce gene expression through a mechanism involving DNA bending and loop formation and often regulate virulence gene expression in cooperation with histone-like proteins (see the chapter on quorum sensing; see also below).

Another large group of transcriptional regulators involved in virulence gene transcription is the LysR family. Apparently, members of this family

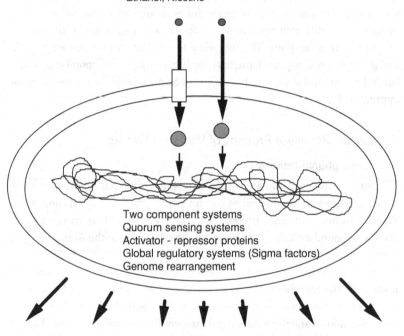

Environmental signals
Stress signals

Temperature (Heat, Cold)
Nutrient Deprivation, Starvation
Oxygen Radicals
Osmolarity
pH (acidic)
Ion content (Mg^{2+}, Ca^{2+}, Fe^{2+})
Ethanol, Nicotine

Two component systems
Quorum sensing systems
Activator - repressor proteins
Global regulatory systems (Sigma factors)
Genome rearrangement

- **Changes of the metabolism - synthesis of special defensive enzymes**

- **Induction of special virulence factors and secretion systems**

- **Variation of the cell surface**

- **Biofilm and microcolony formation**

Figure 2.1. **Signal transduction and regulation of gene expression in pathogens.** Specific virulence genes as well as generally protective genes that contribute to survival and growth of a pathogen within a host are controlled by numerous environmental signals. Signalling molecules may enter the cell spontanously or via specific uptake systems, and then affect the activity of regulator proteins, which in turn activate or repress certain genes. Some transmembrane sensor systems (e.g., in two-component systems) do not transfer a

(cont.)

evolved from a distant ancestor into subfamilies with highly diverse functions found in different prokaryotic genera. They include a number of regulators involved in amino acid biosynthesis, stress responses, chromosome replication and antibiotic resistance. In general, regulators of the LysR family are positive transcriptional regulatory proteins with a negative autoregulatory effect (121). However, several LysR-like proteins also can act as repressors (32, 76) (see the chapter on quorum sensing).

Globally acting regulatory factors which are common in many bacteria can be regulators of virulence factor expression in pathogens. For instance, the cyclic AMP–CRP complex, which is crucial for catabolite repression in Gram-negative bacteria, not only regulates metabolically important genes in response to nutrient availability but also often controls the expression of virulence factors. Since interaction of CRP with DNA depends on the presence of cAMP, transcription of CRP-regulated genes can be controlled by the intracellular cAMP level, which is modulated by nutritional conditions. CRP contains an HTH motif that recognizes a specific CRP-consensus sequence, and activation occurs either directly through contact of the CRP protein with the α-subunit of RNA polymerase (RNAP) or indirectly by affecting DNA topology that facilitates contact with other regulatory proteins and/or RNAP. cAMP–CRP seems to play a crucial role in the virulence of several pathogenic organisms, including *Salmonella*, *Vibrio cholerae* and *E. coli* (75, 126). Given the pleiotropic nature of *crp* mutants it is difficult to know precisely how cAMP–CRP contributes to virulence in a given pathogen. It is always possible that several independent effects are involved. For example, CRP is involved in flagella synthesis of *S. typhimurium*, a function that affects virulence of pathogens particularly during the adhesion and invasion process (86). Furthermore, cAMP–CRP regulates the *ompB* and the *spv* loci, which contribute significantly to the survival of *Salmonella* in mice (38). Interestingly, in *E. coli* a regulatory overlap exists between the cAMP – CRP and the iron-limitation regulon. It has been shown that CRP regulates the transcription of *fur*, the gene encoding the global pleiotropic iron regulatory protein Fur (see above).

Figure 2.1. (*cont.*) signalling molecule, but rather information about its presence into the cell. They may do so by communicating directly (e.g., via conformational alterations and/or chemical modification) or indirectly (e.g., by producing a second messenger molecule like cAMP) with a regulatory protein, which then affects gene expression. Due to cooperation of these signal transduction and regulatory pathways in large regulatory networks as well as the modular nature of bacterial promoter regions, in which several binding sites for regulatory proteins can be combined, extreme regulatory precision and specificity of relevant gene expression patterns are achieved.

Two-Component Regulatory Systems and Phosphorelay Signal Transduction

The control of many virulence genes involves "two-component" regulatory systems which sense certain environmental conditions and transduce this information into the bacterial cell. The intracellular output of a two-component system is usually transcriptional activation of genes that encode factors which allow the cells to cope with the inducing adverse environmental conditions. Bacteria can possess up to several dozen two-component systems with different input signals and output promoter specificities which regulate a wide range of bacterial functions such as chemotaxis, stress responses, or virulence (67, 107). These systems consist of a sensor kinase, which usually spans the inner membrane and an extracellular sensor domain that perceives a given external signal (e.g., certain small molecules or osmolarity). Upon sensing the stimulus, the intracellular histidine kinase domain of the protein (transmitter) undergoes autophosphorylation. The phosphate residue is subsequently transferred to an Asp residue in the N-terminal domain (receiver) of a response regulator protein. This phosphotransfer activates the C-terminal output domain of the response regulator, which can result in activation or repression of target genes (20). Structure and organization of two-component regulatory systems are highly flexible and variable. For instance, the phosphate of a sensor kinase transmitter domain can be serially transferred via a receiver and a hisphosphotransferase (HPT) domain to yet another response regulator (phosphorelay system). The additional components provide more targets for signal integration.

To date, more than 100 two-component systems have been identified in prokaryotes and many of these systems are involved in pathogenesis. One of the best characterized systems in pathogenic bacteria is PhoP/PhoQ that mediates the adaptation of *Salmonella* to Mg^{2+}-limiting environments and regulates numerous cellular functions including virulence (54). Since the Mg^{2+} level inside eukaryotic cells is usually very low, in contrast to the high Mg^{2+} concentration in the extracellular environment, it is believed that *Salmonella* uses PhoP/PhoQ to determine its subcellular location by sensing Mg^{2+}, and thereby ensures that certain virulence genes are expressed inside cells.

General and Global Regulatory Mechanisms of Virulence Genes

Coordination of gene expression is achieved at several levels and includes the grouping of genes for adequate regulation. Virulence genes are often grouped physically, as is the case of large operons encoded in pathogenicity islands, virulence plasmids or lysogenic phages. This allows all the genes to be switched on and off simultaneously. In other cases, genes/operons are dispersed throughout the chromosome or are even located on separate replicons. These diverse genes can be under the control of a common regulatory protein (regulon). Furthermore, virulence genes can belong to several regulons, which is reflected by complex promoter regions that carry binding sites for several regulators. Such genes can respond to more than one stimulus, i.e., at their promoter regions information about different environmental parameters is integrated. Sensing many individual stress signals at a time ensures that essential virulence genes which are needed for survival are only induced at the appropriate site and time during the infection process. These regulatory hierarchies make use of very pleiotropic control mechanisms. The regulators involved are often alternative sigma subunits of RNAP or proteins with global regulatory function as well as nucleoid-structuring functions. Thus, overall or local DNA topology can be involved.

Alternative Sigma Factors

Several bacterial pathogens use alternative sigma factors to control the transcription of large subsets of genes that may include virulence genes. One significant example is the alternative sigma factor σ^S (RpoS) that is induced during entry into stationary phase as well as in response to many environmental stress conditions (see above). σ^S has been shown to be also involved in the control of virulence of a number of pathogens. Initial work with *S. typhimurium* has shown that an *rpoS* mutant exhibits significantly reduced virulence in mice (40, 138). However, the mechanism by which σ^S affects the pathogenicity of *Salmonella* remains incompletely defined. σ^S is required to express a large number of stress proteins essential for resistance against oxidative, acidic and osmotic stress, starvation and DNA damage (see above), all of which are stress conditions the bacteria encounter in host tissue, especially upon uptake into macrophages. Thus, it has been proposed that the ability to survive antimicrobial activity of phagocytes and to resist prolonged deprivation of nutrition may pertain directly to the ability of *Salmonella* to persist and multiply in host tissue (138). In addition, it has been shown that the activity of σ^S rapidly increases after bacterial entry into eukaryotic cells, resulting in the induction of stress and virulence genes responsible for the colonization of deeper tissue. One example is the plasmid-borne *spvABCD* operon known to be required for multiplication within the reticulo-endothelial system and the production of progressive systemic disease (24, 102, 138). However, influence of *rpoS* on *spv* expression does not explain the defects of *rpoS* mutants in earlier stages in the infection. Thus, it has been proposed that other *rpoS*-regulated genes, such as the periplasmic Cu,Zn-superoxide dismutase and a non-heme catalase involved in *Salmonella* resistance against oxidative stress, might also contribute to *Salmonella* virulence (39, 113, 129).

An influence of σ^S on colonization, persistence and survival in host environments has also been reported for other Gram-negative pathogens. σ^S seems to affect quite different stages of the infectious process and the effect of σ^S on pathogenicity varies significantly between microorganisms, even within a species. For instance, in *V. cholerae* C6709 σ^S plays a significant role in colonization of the intestine, whereas no influence on colonization was found in *V. cholerae* biotype O385 and 92A1552 (96, 141). σ^S is also not always required for virulence. It has been reported that σ^S is not involved in *Y. enterocolitica* virulence, although it does contribute to survival under environmental stress and is required for the production of extracellular toxins (69). Moreover, σ^S is not exclusive in its role as regulator of stress-induced resistance genes, since *rpoS* mutant strains can survive certain environmental insults, indicating that additonal uncharacterized stress-dependent regulatory mechanisms also contribute to the survival of Gram-negative microorganisms under stress (136). Interestingly, each pathogen seems to use a very specific combination of general and specific stress-responsive regulatory mechanisms to optimize the expression of its virulence genes according to its specific lifestyle within or outside of the host. Thus, in some pathogens, σ^S may be crucial, whereas in others it may just play a marginal role. σ^S may also be important for survival in a non-human, i.e., environmental host. This has been described for *Legionella pneumophila*, which requires σ^S for survival in *Acanthamoeba*, whereas the role of σ^S for virulence in the human host is still under debate (5, 56).

Sigma B (σ^B) plays an analogous role to σ^S in Gram-positive bacteria and induces a large number of stress proteins in response to specific environmental stimuli, such as heat shock, osmolarity, glucose starvation, oxygen limitation and oxidative stress (53, 59). Since σ^B increases stress survival, it has been speculated that it might be essential for surviving low-pH environments characteristic of the phagolysosomes. In fact, a role in virulence has been reported for *Bacillus anthracis*, the etiological agent of anthrax (44). In contrast, σ^B-directed genes do not appear to be essential for the spread of *L. monocytogenes* to mouse liver and spleen (137). σ^B does not seem to have an effect on the pathogenicity of *S. aureus* in a mouse subcutaneous abscess model (22). However, SarA, a major regulator of a number of representative virulence factors in *S. aureus*, upregulates σ^B expression upon entry into stationary phase, and σ^B functions as a global regulator that induces the expression of several virulence features including hemolysin production and fibronectin-binding (25, 79). As observed with σ^S in Gram-negative bacteria, several avirulent laboratory strains of gram-positives are σ^B-defective. Thus, in both Gram-negative and Gram-positive bacteria, regulation of virulence genes and the response to environmental stresses are closely linked and this connection seems essential for survival and multiplication of the pathogens in their hosts.

Besides the typical stress-associated σ-factors, other alternative σ-factors are also known to be involved in virulence gene expression and stress survival. For instance, the synthesis of the general stress protein GspA of *L. pneumophila* is induced by the heat shock σ-factor (σ^H) during the intracellular infection of macrophages (1). Furthermore, the nitrogen starvation σ-factor (σ^N) of *P. aeruginosa* activates the expression of Type IV fimbriae under nitrogen deprivation. Several studies also reported that the flagella-dependent motility of bacterial pathogens contributes significantly to their pathogenicity and host survival (105). Potential benefits of motility include the ability to translocate to preferred hosts and to access optimal colonization sites, as well as increased efficiency of nutrient acquisition and avoidance of toxic substances. In the past few years, it has become increasingly apparent that σ^F (FliA), which is responsible for the expression of flagella proteins involved in the final stages of flagellar assembly, also regulates virulence genes, such as the type III secretion genes encoded on centrisome 63 of *S. typhimurium* (35). In addition to common sigma factors, pathogens can also synthesize unique alternative sigma factors specialized for the expression of individual virulence factors. One example is AlgU (σ^{22}), which mediates mucoidy in *P. aeruginosa* by inducing the synthesis of the exopolysaccharide alginate, a major virulence determinant expressed during respiratory infections in cystic fibrosis patients. This system is also responsible for enhancing bacterial resistance against heat and oxidative stress and is absolutely required for full virulence (122).

DNA Supercoiling and Histone-Like Proteins

Another global mechanism which changes the genetic program of pathogens for adaptation to host-associated stresses involves alterations in DNA topology. Prokaryotic DNA is arranged in the nucleoid in a negatively supercoiled (underwound) state, which favors reactions that depend on strand separation, including transcription. It has been known for several years that environmental parameters which are likely to be experienced during infection, such as changes in nutrient and oxygen availability, temperature and osmolarity can affect overall DNA supercoiling. This in turn can influence the transcriptional activity of many promoters, including those of stress and virulence genes (34). For instance, the invasion genes of *S. flexneri* are transcribed at 37° C, but not at 25° C, by an AraC-like activator protein

(VirF). Mutations in genes encoding proteins known to alter DNA supercoiling, such as topoisomerases and histone-like proteins (see below), induce virulence gene expression even at 25° C, suggesting that temperature control of the promoter is linked to changes in DNA topology (135).

H-NS and Hha/YmoA are both small nucleoid-associated histone-like proteins which globally repress many genes including those encoding virulence functions (4, 21, 27, 37). Many of the H-NS- and Hha-controlled promoters are also subject to regulation by changes of DNA supercoiling and, in some cases, these proteins may repress promoter activity by locally modifying DNA topology. Alternatively, the large oligomer-forming protein H-NS can silence genes by establishing large nucleoprotein complexes covering entire promoter regions (29). This repression is modulated by environmental parameters, such as temperature, osmolarity, pH and oxygen tension. Often, negative regulation by YmoA/Hha or H-NS is counteracted by transcriptional activator proteins of the AraC family. Thus, H-NS-repressed genes seem to be activated either by competitive displacement by a specific activator protein or H-NS displacement due to an alteration in DNA topology induced by environmental signals (4, 117). Also the histone-like DNA-bending proteins Fis, IHF and Lrp can act as global modulators for environmentally controlled virulence factors. The expression of these nucleoid-associated proteins is itself controlled by environmental conditions. Often, several histone-like proteins act together with transcriptional regulators to determine the final level of virulence gene transcription. For instance, the *spv* virulence genes of *S. typhimurium* are under positive control of σ^S, SpvR, a LysR-like activator protein and the histone-like protein IHF. In addition, the *spv* genes are repressed by H-NS and Lrp (94, 114). Thus, all these factors together probably generate large promoter-specific nucleoprotein complexes. The modularity of these systems produces enormous flexibility with respect to environmental signal integration and regulatory fine-tuning.

CONCLUSIONS

The regulation of bacterial pathogenicity, and the control of expression of specific virulence genes in particular, is highly interwoven with the general cellular regulatory network of bacteria. This network is organised in a highly hierarchical and modular fashion that provides extreme flexibility and precision of control. Very globally acting master-regulators, e.g., sigma factors, cAMP–CRP or certain two-component systems, can cooperate with more specific regulators in seemingly limitless combinations. These regulatory factors perform their functions in response to numerous environmental signals reflecting the fact that bacteria are exposed to ever-changing stress conditions. In particular, essential nutrients tend to be scarce within natural environments, both external and within specific host organisms. Thus, pathogens, just like any other bacteria, spend most of their life slowly growing or even in stationary phase. As a consequence, virulence genes are not only controlled in response to the very diverse stress conditions that the pathogen typically encounters in the course of its journey through the host but also tend to be stationary-phase–inducible. Moreover, general and specific stress resistance mechanisms are of utmost importance for survival of the pathogen in a host that tries hard to maximize stress for the pathogen in its attempt to get rid of the latter. From the perspective of the pathogen, however, a host is just a complex, potentially rich but also potentially life-threatening environment, in which it has not only to survive but also to find or even establish a niche where it can grow and divide. The production of virulence factors and their complex regulation in response to environmental and intra-host conditions just serve to find such niches and make them as productive as possible for multiplication at the lowest cost for the pathogen.

ACKNOWLEDGEMENTS

Research in the laboratories of R.H.-A. is financially supported by the Deutsche Forschungsgemeinschaft (Gottfried–Wilhelm–Leibniz program, awarded to R.H.-A.; grant DE616/2-1, awarded to P.D.), the State of Baden–Württemberg (Landesforschungspreis, awarded to R.H.-A.), and the Fonds der Chemischen Industrie.

REFERENCES

1. Abu Kwaik, Y., Gao, L. Y., Harb, O. S., and Stone, B. J. 1997. Transcriptional regulation of the macrophage-induced gene (gspA) of Legionella pneumophila and phenotypic characterization of a null mutant. Mol. Microbiol. 24(3):629–642.
2. Aldea, M., Garrido, T., Hernández-Chico, C., Vicente, M., and Kushner, S. R. 1989. Induction of a growth-phase-dependent promoter triggers transcription of bolA, an Escherichia coli morphogene. EMBO J. 8:3923–3931.
3. Arnqvist, A., Olsén, A., and Normark, S. 1994. σ^S-dependent growth-phase induction of the csgBA promoter in Escherichia coli can be achieved in vivo by σ^{70} in the absence of the nucleoid-associated protein H-NS. Mol. Microbiol. 13:1021–1032.

4. Atlung, T., and Ingmer, H. 1997. H-NS: A modulator of environmentally regulated gene expression. Mol. Microbiol. **24**(1):7–17.

5. Bachman, M. A., and Swanson, M. S. 2001. RpoS co-operates with other factors to induce *Legionella pneumophila* virulence in the stationary phase. Mol. Microbiol. **40**:1201–1214.

6. Balandina, A., Claret, L., Hengge-Aronis, R., and Rouvière-Yaniv, J. 2001. The *Escherichia coli* histone-like protein HU regulates *rpoS* translation. Mol. Microbiol. **39**:1069–1079.

7. Barth, M., Marschall, C., Muffler, A., Fischer, D., and Hengge-Aronis, R. 1995. A role for the histone-like protein H-NS in growth phase-dependent and osmotic regulation of σ^S and many σ^S-dependent genes in *Escherichia coli*. J. Bacteriol. **177**:3455–3464.

8. Bearson, S. M. D., Benjamin, W. H., Jr., Swords, W. E., and Foster, J. W. 1996. Acid shock induction of RpoS is mediated by the mouse virulence gene *mviA* of *Salmonella typhimurium*. J. Bacteriol. **178**:2572–2579.

9. Becker, G., and Hengge-Aronis, R. 2001. What makes an *Escherichia coli* promoter σ^S-dependent? Role of the $-13/-14$ nucleotide promoter positions and region 2.5 of σ^S. Mol. Microbiol. **39**:1153–1165.

10. Becker, G., Klauck, E., and Hengge-Aronis, R. 1999. Regulation of RpoS proteolysis in *Escherichia coli*: The response regulator RssB is a recognition factor that interacts with the turnover element in RpoS. Proc. Natl. Acad. Sci. USA **96**:6439–6444.

11. Bleves, S., and Cornelis, G. R. 2000. How to survive in the host: The *Yersinia* lesson. Microbes Infect. **2**(12):1451–1460.

12. Booth, I. R. 1982. The regulation of intracellular pH in bacteria. In J. C. Derek and G. Gardner (Eds.), *Bacterial Responses to pH*, pp. 19–27. John Wiley & Sons, Ltd., Chichester, England.

13. Bouvier, J., Gordia, S., Kampmann, G., Lange, R., Hengge-Aronis, R., and Gutierrez, C. 1998. Interplay between global regulators of *Escherichia coli*: Effect of RpoS, H-NS and Lrp on transcription of the gene *osmC*. Mol. Microbiol. **28**:971–980.

14. Boyd, J., Oza, M. N., and Murphy, J. R. 1990. Molecular cloning and DNA sequence analysis of a diphtheria toxin iron-dependent regulatory element (*dtxR*) from *Corynebacterium diphtheriae*. Proc. Natl. Acad. Sci. USA **87**(15):5968–5972.

15. Braun, V. 2000. The art of keeping low and high iron concentration in balance. In G. Storz and R. Hengge-Aronis (Eds.), *Bacterial Stress Responses*, pp. 275–278. American Society of Microbiology, Washington, D.C.

16. Braun, V. 2001. Iron uptake mechanisms and their regulation in pathogenic bacteria. Int. J. Med. Microbiol. **291**(2):67–79.

17. Brown, L., and Elliott, T. 1997. Mutations that increase expression of the *rpoS* gene and decrease its dependence on *hfq* function in *Salmonella typhimurium*. J. Bacteriol. **179**:656–662.

18. Buchanan, R. L., and Edelson, S. G. 1996. Culturing enterohemorrhagic *Escherichia coli* in the presence and absence of glucose as a simple means of evaluating the acid tolerance of stationary-phase cells. Appl. Environ. Microbiol. **62**(11):4009–4013.

19. Buchmeier, N. A., Libby, S. J., Xu, Y., Loewen, P. C., Switala, J., Guiney, D. G., and Fang, F. C. 1995. DNA repair is more important than catalase for *Salmonella* virulence in mice. J. Clin. Invest. **95**(3):1047–1053.

20. Buckler, D. R., Anand, G. S., and Stock, A. M. 2000. Response-regulator phosphorylation and activation: A two-way street? Trends Microbiol. **8**(4):153–156.

21. Bustamante, V., Santana, F., Calva, E., and Puente, J. 2001. Transcriptional regulation of type III secretion genes in enteropathogenic *Escherichia coli*: Ler antagonizes H-NS-dependent repression. Mol. Microbiol. **39**(3):664–678.

22. Chan, P. F., Foster, S. J., Ingham, E., and Clements, M. O. 1998. The *Staphylococcus aureus* alternative sigma factor sigmaB controls the environmental stress response but not starvation survival or pathogenicity in a mouse abscess model. J. Bacteriol. **180**(23):6082–6089.

23. Chang, Y.-Y., Wang, A.-Y., and Cronan, J. E., Jr. 1994. Expression of *Escherichia coli* pyruvate oxidase (PoxB) depends on the sigma factor encoded by the *rpoS* (*katF*) gene. Mol. Microbiol. **11**:1019–1028.

24. Chen, C. Y., Eckmann, L., Libby, S. J., Fang, F. C., Okamoto, S., Kagnoff, M. F., Fierer, J., and Guiney, D. G. 1996. Expression of *Salmonella typhimurium rpoS* and *rpoS*-dependent genes in the intracellular environment of eukaryotic cells. Infect. Immun. **64**(11):4739–4743.

25. Cheung, A. L., Chien, Y. T., and Bayer, A. S. 1999. Hyperproduction of alpha-hemolysin in a *sigB* mutant is associated with elevated SarA expression in *Staphylococcus aureus*. Infect. Immun. **67**(3):1331–1337.

26. Colland, F., Barth, M., Hengge-Aronis, R., and Kolb, A. 2000. Sigma factor selectivity of *Escherichia coli* RNA polymerase: A role for CRP. IHF and Lrp transcription factors. EMBO J. **19**:3028–3037.

27. Cornelis, G. R. 1993. Role of the transcription activator *virF* and the histone-like protein YmoA in the thermoregulation of virulence functions in *yersiniae*. Zentralbl. Bakteriol. **278**(2–3):149–164.

28. Cornelis, G. R. 2000. Type III secretion: A bacterial device for close combat with cells of their eukaryotic host. Philos. Trans. R. Soc. Lond. B Biol. Sci. **355**(1397):681–693.

29. Dame, R. T., Wyman, C., Wurm, R., Wagner, R., and Goosen, N. 2001. Structural basis for H-NS mediated trapping of RNA polymerase in the open initiation complex at the *rrnB* P1. J. Biol. Chem. **277**(3): 2146–2150.

30. de Kievit, T. R., and Iglewski, B. H. 2000. Bacterial quorum sensing in pathogenic relationships. Infect. Immun. **68**(9):4839–4849.

31. de Lorenzo, V., Giovannini, F., Herrero, M., and Neilands, J. B. 1988. Metal ion regulation of gene expression. Fur repressor-operator interaction at the promoter region of the aerobactin system of pColV-K30. J. Mol. Biol. **203**(4):875–884.

32. Deghmane, A. E., Petit, S., Topilko, A., Pereira, Y., Giorgini, D., Larribe, M., and Taha, M. K. 2000. Intimate adhesion of *Neisseria meningitidis* to human epithelial cells is under the control of the *crgA* gene, a novel LysR-type transcriptional regulator. EMBO J. **19**(5):1068–1078.

33. Demple, B. 1996. Redox signaling and gene control in the *Escherichia coli soxRS* oxidative stress regulon – a review. Gene. **179**(1):53–57.

34. Dorman, C. J. 1991. DNA supercoiling and environmental regulation of gene expression in pathogenic bacteria. Infect. Immun. **59**(3): 745–749.

35. Eichelberg, K., and Galan, J. E. 2000. The flagellar sigma factor FliA (sigma(28)) regulates the expression of *Salmonella* genes associated with the centisome 63 type III secretion system. Infect. Immun. **68**(5):2735–2743.

36. Espinosa-Urgel, M., Chamizo, C., and Tormo, A. 1996. A consensus structure for σ^S-dependent promoters. Mol. Microbiol. **21**:657–659.

37. Fahlen, T. F., Mathur, N., and Jones, B. D. 2000. Identification and characterization of mutants with increased expression of *hilA*, the invasion gene transcriptional activator of *Salmonella typhimurium*. FEMS Immunol. Med. Microbiol. **28**(1):25–35.

38. Fang, F. C., Chen, C. Y., Guiney, D. G., and Xu, Y. 1996. Identification of sigma S-regulated genes in *Salmonella typhimurium*: Complementary regulatory interactions between sigma S and cyclic AMP receptor protein. J. Bacteriol. **178**(17):5112–5120.

39. Fang, F. C., DeGroote, M. A., Foster, J. W., Baumler, A. J., Ochsner, U., Testerman, T., Bearson, S., Giard, J. C., Xu, Y., Campbell, G., and Laessig, T. 1999. Virulent *Salmonella typhimurium* has two periplasmic Cu, Zn-superoxide dismutases. Proc. Natl. Acad. Sci. USA **96**(13):7502–7507.

40. Fang, F. C., Libby, S. J., Buchmeier, N. A., Loewen, P. C., Switala, J., Harwood, J., and Guiney, D. G. 1992. The alternative sigma factor katF (*rpoS*) regulates *Salmonella* virulence. Proc. Natl. Acad. Sci. USA **89**(24): 11,978–11,982.

41. Farr, S. B., and Kogoma, T., 1991. Oxidative stress responses in *Escherichia coli* and *Salmonella typhimurium*. Microbiol. Rev. **55**(4):561–585.

42. Fischer, D., Teich, A., Neubauer, P., and Hengge-Aronis, R. 1998. The general stress sigma factor σ^S of *Escherichia coli* is induced during diauxic shift from glucose to lactose. J. Bacteriol. **180**:6203–6206.

43. Foster, J. W. 2000. Microbial responses to acidic stress. In G. Storz and R. Hengge-Aronis (Eds.), *Bacterial Stress Responses*, pp. 99–115. American Society of Microbiology, Washington, D.C.

44. Fouet, A., Namy, O., and Lambert, G. 2000. Characterization of the operon encoding the alternative sigma(B) factor from *Bacillus anthracis* and its role in virulence. J. Bacteriol. **182**(18):5036–5045.

45. Fuchs, S., Muhldorfer, I., Donohue-Rolfe, A., Kerenyi, M., Emody, L., Alexiev, R., Nenkov, P., and Hacker, J. 1999. Influence of RecA on in vivo virulence and Shiga toxin 2 production in *Escherichia coli* pathogens. Microb. Pathog. **27**(1):13–23.

46. Fuqua, C., Winans, S. C., and Greenberg, E. P. 1996. Census and consensus in bacterial ecosystems: The LuxR-LuxI family of quorum-sensing transcriptional regulators. Annu. Rev. Microbiol. **50**:727–751.

47. Fuqua, W. C., Winans, S. C., and Greenberg, E. P. 1994. Quorum sensing in bacteria: The *luxR-luxI* family of cell density-responsive transcriptional regulators. J. Bacteriol. **176**:269–275.

48. Galan, J. E. 1996. Molecular genetic bases of *Salmonella* entry into host cells. Mol. Microbiol. **20**(2):263–271.

49. Gallegos, M. T., Schleif, R., Bairoch, A., Hofmann, K., and Ramos, J. L. 1997. Arac/XylS family of transcriptional regulators. Microbiol. Mol. Biol. Rev. **61**(4):393–410.

50. Garcia del Portillo, F., Foster, J. W., and Finlay, B. B. 1993. Role of acid tolerance response genes in *Salmonella typhimurium* virulence. Infect. Immun. **61**(10):4489–4492.

51. Gentry, D. R., Hernandez, V. J., Nguyen, L. H., Jensen, D. B., and Cashel, M. 1993. Synthesis of the stationary-phase sigma factor σ^S is positively regulated by ppGpp. J. Bacteriol. **175**:7982–7989.

52. Germer, J., Becker, G., Metzner, M., and Hengge-Aronis, R. 2001. Role of activator site position and a distal UP-element half-site for sigma factor selectivity at a CRP/H-NS activated σ^S-dependent promoter in *Escherichia coli*. Mol. Microbiol. **41**:705–716.

53. Gertz, S., Engelmann, S., Schmid, R., Ziebandt, A. K., Tischer, K., Scharf, C., Hacker, J., and Hecker, M. 2000. Characterization of the sigma(B) regulon in *Staphylococcus aureus*. J. Bacteriol. **182**(24):6983–6991.

54. Groisman, E. A. 2001. The pleiotropic two-component regulatory system PhoP-PhoQ. J. Bacteriol. **183**(6):1835–1842.

55. Gutierrez, C., Barondess, J., Manoil, C., and Beckwith, J. 1987. The use of transposon Tn*phoA* to detect genes for cell envelope proteins subject to a common regulatory stimulus. J. Mol. Biol. **195**:289–297.

56. Hales, L. M., and Shuman, H. A. 1999. *Legionella pneumophila rpoS* is required for growth within *Acanthamoeba castellanii*. J. Bacteriol. **181**:4879–4889.

57. Hantke, K., and Braun, V. 2000. Microbial responses to acidic stress. In G. Storz and R. Hengge-Aronis (Eds.), *Bacterial Stress Responses*, pp. 275–288. American Society of Microbiology, Washington, D.C.

58. Hastings, J. W., and Greenberg, E. P. 1999. Quorum sensing: The explanation of a curious phenomenon reveals a common characteristic of bacteria. J. Bacteriol. **181**:2667–2668.

59. Hecker, M., Schumann, W., and Volker, U. 1996. Heat-shock and general stress response in *Bacillus subtilis*. Mol. Microbiol. **19**(3):417–428.

60. Heesemann, J., Hantke, K., Vocke, T., Saken, E., Rakin, A., Stojiljkovic, I., and Berner, R. 1993. Virulence of *Yersinia enterocolitica* is closely associated with siderophore production, expression of an iron-repressible outer membrane polypeptide of 65,000 Da and pesticin sensitivity. Mol. Microbiol. **8**(2):397–408.

61. Henderson, B., Poole, S., and Wilson, M. 1996. Bacterial modulins: A novel class of virulence factors which cause host tissue pathology by inducing cytokine synthesis. Microbiol. Rev. **60**(2):316–341.

62. Hengge-Aronis, R. 2000. The general stress response in *Escherichia coli*. In G. Storz and R. Hengge-Aronis (Eds.), *Bacterial Stress Responses*, pp. 161–178. ASM Press, Washington, D.C.

63. Hengge-Aronis, R. 1999. Interplay of global regulators in the general stress response of *Escherichia coli*. Curr. Opin. Microbiol. **2**(2):148–152.

64. Hengge-Aronis, R. 2002. Recent insights into the general stress response regulatory network in *Escherichia coli*. J. Mol. Microbiol. Biotech. **4**(3):341–346.

65. Hengge-Aronis, R. 1996. Regulation of gene expression during entry into stationary phase. In F. C. Neidhardt (Ed.), Escherichia coli *and* Salmonella typhimurium: *Cellular and Molecular Biology*, pp. 1497–1512. ASM Press, Washington, D.C.

65a. Hengge-Aronis, R. 2002. Signal transduction and regulatory mechanisms involved in control of the σ^S (RpoS) subunit of RNA polymerase. Microbiol. Molec. Biol. Rev. **66**:373–395.

66. Hengge-Aronis, R. 1993. Survival of hunger and stress: The role of *rpoS* in stationary phase gene regulation in *Escherichia coli*. Cell **72**:165–168.

67. Hoch, J. A. 2000. Two-component and phosphorelay signal transduction. Curr. Opin. Microbiol. 3(2):165–170.

68. Hueck, C. J. 1998. Type III protein secretion systems in bacterial pathogens of animals and plants. Microbiol. Mol. Biol. Rev. 62(2):379–433.

69. Iriarte, M., Stainier, I., and Cornelis, G. R. 1995. The *rpoS* gene from *Yersinia enterocolitica* and its influence on expression of virulence factors. Infect. Immun. 63(5):1840–1847.

70. Isberg, R. R. 2000. Identification and analysis of proteins expressed by bacterial pathogens in response to host tissues. In G. Storz and R. Hengge-Aronis (Eds.), *Bacterial Stress Responses*, pp. 289–303. ASM Press, Washington, D.C.

71. Isberg, R. R., Hamburger, Z., and Dersch, P. 2000. Signaling and invasin-promoted uptake via integrin receptors. Microbes Infect. 2(7):793–801.

72. Isberg, R. R., and Leong, J. M. 1990. Multiple beta 1 chain integrins are receptors for invasin, a protein that promotes bacterial penetration into mammalian cells. Cell 60(5):861–871.

73. Jenkins, D. E., Chaisson, S. A., and Matin, A. 1990. Starvation-induced cross-protection against osmotic challenge in *Escherichia coli*. J. Bacteriol. 172:2779–2781.

74. Jenkins, D. E., Schultz, J. E., and Matin, A. 1988. Starvation-induced cross-protection against heat or H_2O_2 challenge in *Escherichia coli*. J. Bacteriol. 170:3910–3914.

75. Kennedy, M. J., Yancey, R. J., Jr., Sanchez, M. S., Rzepkowski, R. A., Kelly, S. M., and Curtiss, R., III. 1999. Attenuation and immunogenicity of Deltacya Deltacrp derivatives of *Salmonella choleraesuis* in pigs. Infect. Immun. 67(9):4628–4636.

76. Klauck, E., Böhringer, J., and Hengge-Aronis, R. 1997. The LysR-like regulator LeuO in *Escherichia coli* is involved in the translational regulation of *rpoS* by affecting the expression of the small regulatory DsrA-RNA. Mol. Microbiol. 25:559–569.

77. Klauck, E., Lingnau, M., and Hengge-Aronis, R. 2001. Role of the response regulator RssB in σ^S recognition and initiation of σ^S proteolysis in *Escherichia coli*. Mol. Microbiol. 40:1381–1390.

78. Kolter, R., Siegele, D. A., and Tormo, A. 1993. The stationary phase of the bacterial life cycle. Annu. Rev. Microbiol. 47:855–874.

79. Kullik, I., Giachino, P., and Fuchs, T. 1998. Deletion of the alternative sigma factor sigmaB in *Staphylococcus aureus* reveals its function as a global regulator of virulence genes. J. Bacteriol. 180(18):4814–4820.

80. Lange, R., Fischer, D., and Hengge-Aronis, R. 1995. Identification of transcriptional start sites and the role of ppGpp in the expression of *rpoS*, the

structural gene for the σ^S subunit of RNA-polymerase in *Escherichia coli*. J. Bacteriol. **177**:4676–4680.

81. Lange, R., and Hengge-Aronis, R. 1994. The cellular concentration of the σ^S subunit of RNA-polymerase in *Escherichia coli* is controlled at the levels of transcription, translation and protein stability. Genes Dev. **8**:1600–1612.

82. Lange, R., and Hengge-Aronis, R. 1991. Growth phase-regulated expression of *bolA* and morphology of stationary phase *Escherichia coli* cells is controlled by the novel sigma factor σ^S (*rpoS*). J. Bacteriol. **173**:4474–4481.

83. Lange, R., and Hengge-Aronis, R. 1991. Identification of a central regulator of stationary-phase gene expression in *Escherichia coli*. Mol. Microbiol. **5**:49–59.

84. Lease, R. A., Cusick, M. E., and Belfort, M. 1998. Riboregulation in *Escherichia coli*: DsrA RNA acts by RNA:RNA interaction at multiple loci. Proc. Natl. Acad. Sci. USA **95**:12,456–12,461.

85. Lee, C. A. 1997. Type III secretion systems: Machines to deliver bacterial proteins into eukaryotic cells? Trends Microbiol. **5**(4):148–156.

86. Lee, C. A., Jones, B. D., and Falkow, S. 1992. Identification of a *Salmonella typhimurium* invasion locus by selection for hyperinvasive mutants. Proc. Natl. Acad. Sci. USA **89**(5):1847–1851.

87. Lee, I. S., Lin, J., Hall, H. K., Bearson, B., and Foster, J. W. 1995. The stationary-phase sigma factor σ^S (RpoS) is required for a sustained acid tolerance response in virulent *Salmonella typhimurium*. Mol. Microbiol. **17**:155–167.

88. Loewen, P. C., and Hengge-Aronis, R. 1994. The role of the sigma factor σ^S (KatF) in bacterial global regulation. Annu. Rev. Microbiol. **48**:53–80.

89. Loewen, P. C., Hu, B., Strutinsky, J., and Sparling, R. 1998. Regulation in the *rpoS* regulon of *Escherichia coli*. Can. J. Microbiol. **44**:707–717.

90. Majdalani, N., Chen, S., Murrow, J., St. John, K., and Gottesman, S. 2001. Regulation of RpoS by a novel small RNA: The characterization of RprA. Mol. Microbiol. **39**:1382–1394.

91. Majdalani, N., Cunning, C., Sledjeski, D., Elliott, T., and Gottesman, S. 1998. DsrA RNA regulates translation of RpoS message by an anti-antisense mechanism, independent of its action as an antisilencer of transcription. Proc. Natl. Acad. Sci. USA **95**:12,462–12,467.

92. Marron, L., Emerson, N., Gahan, C. G., and Hill, C. 1997. A mutant of *Listeria monocytogenes* LO28 unable to induce an acid tolerance response displays diminished virulence in a murine model. Appl. Environ. Microbiol. **63**:4945–4947.

93. Marshall, B. J., Barrett, L. J., Prakash, C., McCallum, R. W., and Guerrant, R. L. 1990. Urea protects *Helicobacter (Campylobacter) pylori* from the bactericidal effect of acid. Gastroenterology **99**(3):697–702.

94. Marshall, D. G., Sheehan, B. J., and Dorman, C. J. 1999. A role for the leucine-responsive regulatory protein and integration host factor in the regulation of the Salmonella plasmid virulence (*spv*) locus in *Salmonella typhimurium*. Mol. Microbiol. **34**(1):134–145.

95. McCann, M. P., Kidwell, J. P., and Matin, A. 1991. The putative σ factor KatF has a central role in development of starvation-mediated general resistance in *Escherichia coli*. J. Bacteriol. **173**:4188–4194.

96. Merrell, D. S., Tischler, A. D., Lee, S. H., and Camilli, A. 2000. *Vibrio cholerae* requires *rpoS* for efficient intestinal colonization. Infect. Immun. **68**(12):6691–6696.

97. Miller, J. F., Mekalanos, J. J., and Falkow, S. 1989. Coordinate regulation and sensory transduction in the control of bacterial virulence. Science **243**(4893):916–922.

98. Moxon, E. R., and Kroll, J. S. 1990. The role of bacterial polysaccharide capsules as virulence factors. Curr. Top. Microbiol. Immunol. **150**:65–85.

99. Muffler, A., Fischer, D., Altuvia, S., Storz, G., and Hengge-Aronis, R. 1996. The response regulator RssB controls stability of the σ^S subunit of RNA polymerase in *Escherichia coli*. EMBO J. **15**:1333–1339.

100. Muffler, A., Fischer, D., and Hengge-Aronis, R. 1996. The RNA-binding protein HF-I, known as a host factor for phage Qβ RNA replication, is essential for the translational regulation of *rpoS* in *Escherichia coli*. Genes Dev. **10**:1143–1151.

101. Nguyen, L. H., Jensen, D. B., Thompson, N. E., Gentry, D. R., and Burgess, R. R. 1993. In vitro functional characterization of overproduced *Escherichia coli katF/rpoS* gene product. Biochemistry **32**:11,112–11,117.

102. Norel, F., Robbe-Saule, V., Popoff, M. Y., and Coynault, C. 1992. The putative sigma factor KatF (RpoS) is required for the transcription of the *Salmonella typhimurium* virulence gene *spvB* in *Escherichia coli*. FEMS Microbiol. Lett. **78**(2–3):271–276.

103. Nunoshiba, T., DeRojas-Walker, T., Tannenbaum, S. R., and Demple, B. 1995. Roles of nitric oxide in inducible resistance of *Escherichia coli* to activated murine macrophages. Infect. Immun. **63**(3):794–798.

104. Oelschlaeger, T. A. 2001. Adhesins as invasins. Int. J. Med. Microbiol. **291**(1):7–14.

105. Ottemann, K. M., and Miller, J. F. 1997. Roles for motility in bacterial-host interactions. Mol. Microbiol. **24**(6):1109–1117.

106. Perez-Perez, G. I., Olivares, A. Z., Cover, T. L., and Blaser, M. J. 1992. Characteristics of *Helicobacter pylori* variants selected for urease deficiency. Infect. Immun. **60**(9):3658–3663.

107. Perraud, A. L., Weiss, V., and Gross, R. 1999. Signalling pathways in two-component phosphorelay systems. Trends Microbiol. **7**(3):115–120.

108. Pratt, L. A., and Silhavy, T. J. 1996. The response regulator, SprE, controls the stability of RpoS. Proc. Natl. Acad. Sci. USA **93**:2488–2492.

109. Price, C. W. 2000. Protective function and regulation of the general stress response in *Bacillus subtilis* and related gram-positive bacteria. In G. Storz and R. Hengge-Aronis (Eds.), *Bacterial Stress Responses*, pp. 179–197. ASM Press, Washington, D.C.

110. Prince, R. W., Storey, D. G., Vasil, A. I., and Vasil, M. L. 1991. Regulation of toxA and *regA* by the *Escherichia coli fur* gene and identification of a Fur homologue in *Pseudomonas aeruginosa* PA103 and PA01. Mol. Microbiol. **5**(11):2823–2831.

110a. Pruteanu, M., and Hengge-Aronis, R. 2002. The cellular level of the recognition factor RssB is rate-limiting for σ^S proteolysis implications for RssB regulation and signal transduction in σ^S turnover in Eschesichia Coli. Mol. Microbiol. **45**:1701–1713.

111. Repoila, F., and Gottesman, S. 2001. Signal transduction cascade for regulation of *rpoS*: Temperature regulation of *dsrA*. J. Bacteriol. **183**:4012–4023.

112. Riesenberg-Wilmes, M. R., Bearson, B., Foster, J. W., and Curtis, R., III. 1996. Role of the acid tolerance response in virulence of *Salmonella typhimurium*. Infect. Immun. **64**(4):1085–1092.

113. Robbe-Saule, V., Coynault, C., Ibanez-Ruiz, M., Hermant, D., and Norel, F. 2001. Identification of a non-haem catalase in *Salmonella* and its regulation by RpoS (sigmaS). Mol. Microbiol. **39**(6):1533–1545.

114. Robbe-Saule, V., Schaeffer, F., Kowarz, L., and Norel, F. 1997. Relationships between H-NS, sigma S, SpvR and growth phase in the control of *spvR*, the regulatory gene of the *Salmonella* plasmid virulence operon. Mol. Gen. Genet. **256**(4):333–347.

115. Robertson, B. D., and Meyer, T. F. 1992. Genetic variation in pathogenic bacteria. Trends Genet. **8**(12):422–427.

116. Roggenkamp, A., Bittner, T., Leitritz, L., Sing, A., and Heesemann, J. 1997. Contribution of the Mn-cofactored superoxide dismutase (SodA) to the virulence of *Yersinia enterocolitica* serotype O8. Infect. Immun. **65**(11):4705–4710.

117. Rohde, J. R., Fox, J. M., and Minnich, S. A. 1994. Thermoregulation in

Yersinia enterocolitica is coincident with changes in DNA supercoiling. Mol. Microbiol. **12**(2):187–199.

118. Römling, U., Bian, Z., Hammar, M., Sierralta, W. D., and Normark, S. 1998. Curli fibers are highly conserved between *Salmonella typhimurium* and *Escherichia coli* with respect to operon structure and regulation. J. Bacteriol. **180**:722–731.

119. Römling, U., Sierralta, W. D., Eriksson, K., and Normark, S. 1998. Multicellular and aggregative behaviour of *Salmonella typhimurum* strains is controlled by mutations in the *agfD* promoter. Mol. Microbiol. **28**:249–264.

120. Santos, J. M., Freire, P., Vicente, M., and Arraiano, C. M. 1999. The stationary-phase morphogene *bolA* from *Escherichia coli* is induced by stress during early stages of growth. Mol. Microbiol. **32**:789–798.

121. Schell, M. A. 1993. Molecular biology of the LysR family of transcriptional regulators. Annu. Rev. Microbiol. **47**:597–626.

122. Schurr, M. J., Yu, H., Boucher, J. C., Hibler, N. S., and Deretic, V. 1995. Multiple promoters and induction by heat shock of the gene encoding the alternative sigma factor AlgU (sigma E) which controls mucoidy in cystic fibrosis isolates of *Pseudomonas aeruginosa*. J. Bacteriol. **177**(19):5670–5679.

123. Schweder, T., Lee, K.-H., Lomovskaya, O., and Matin, A. 1996. Regulation of *Escherichia coli* starvation sigma factor (σ^S) by ClpXP protease. J. Bacteriol. **178**:470–476.

124. Setlow, P. 2000. Resistance of bacterial spores. In G. Storz and R. Hengge-Aronis (Eds.), *Bacterial Stress Responses*, pp. 217–230. ASM Press, Washington, D.C.

125. Seyler, R. W., Jr., Olson, J. W., and Maier, R. J. 2001. Superoxide dismutase-deficient mutants of *Helicobacter pylori* are hypersensitive to oxidative stress and defective in host colonization. Infect. Immun. **69**(6):4034–4040.

126. Skorupski, K., and Taylor, R. K. 1997. Sequence and functional analysis of the gene encoding *Vibrio cholerae* cAMP receptor protein. Gene **198**(1–2): 297–303.

127. Sledjeski, D. D., Gupta, A., and Gottesman, S. 1996. The small RNA, DsrA, is essential for the low temperature expression of RpoS during exponential growth in *E. coli*. EMBO J. **15**:3993–4000.

128. Sonenshein, A. L. 2000. Bacterial sporulation: A response to environmental signals. In G. Storz and R. Hengge-Aronis (Eds.), *Bacterial Stress Responses*, pp. 199–215. ASM Press, Washington, D.C.

129. Spector, M. P., Garcia del Portillo, F., Bearson, S. M., Mahmud, A., Magut, M., Finlay, B. B., Dougan, G., Foster, J. W., and Pallen, M. J. 1999. The

rpoS-dependent starvation-stress response locus *stiA* encodes a nitrate reductase (narZYWV) required for carbon-starvation-inducible thermotolerance and acid tolerance in *Salmonella typhimurium*. Microbiology 145(Pt 11):3035–3045.

130. Storz, G., and Hengge-Aronis, R. (Eds.). 2000. *Bacterial Stress Responses*. ASM Press, Washington, D.C.

131. Storz, G., and Imlay, J. A. 1999. Oxidative stress. Curr. Opin. Microbiol. 2:188–194.

132. Sutherland, I. W. 2001. The biofilm matrix – an immobilized but dynamic microbial environment. Trends Microbiol. 9(5):222–227.

133. Tanaka, K., Kusano, S., Fujita, N., Ishihama, A., and Takahashi, H. 1995. Promoter determinants for *Escherichia coli* RNA polymerase holoenzyme containing σ^{38} (the *rpoS* gene product). Nucl. Acids Res. 23:827–834.

134. Tanaka, K., Takayanagi, Y., Fujita, N., Ishihama, A., and Takahashi, H. 1993. Heterogeneity of the principal sigma factor in *Escherichia coli*: The *rpoS* gene product, σ^{38}, is a second principal sigma factor of RNA polymerase in stationary phase *Escherichia coli*. Proc. Natl. Acad. Sci. USA 90:3511–3515.

135. Tobe, T., Yoshikawa, M., and Sasakawa, C. 1994. Deregulation of temperature-dependent transcription of the invasion regulatory gene, *virB*, in *Shigella* by *rho* mutation. Mol. Microbiol. 12(2):267–276.

136. Wang, Y., and Kim, K. S. 2000. Effect of *rpoS* mutations on stress-resistance and invasion of brain microvascular endothelial cells in *Escherichia coli* K1. FEMS Microbiol. Lett. 182(2):241–247.

137. Wiedmann, M., Arvik, T. J., Hurley, R. J., and Boor, K. J. 1998. General stress transcription factor sigmaB and its role in acid tolerance and virulence of *Listeria monocytogenes*. J. Bacteriol. 180(14):3650–3656.

138. Wilmes-Riesenberg, M. R., Foster, J. W., and Curtiss, R., III. 1997. An altered rpoS allele contributes to the avirulence of *Salmonella typhimurium* LT2. Infect. Immun. 65(1):203–210.

139. Winans, S. C., and Zhu, J. 2000. The role of cell-cell communication in confronting the limitations and opportunities of high population densities. In G. Storz and R. Hengge-Aronis (Eds.), *Bacterial Stress Responses*, pp. 261–272. ASM Press, Washington, D.C.

140. Yamashino, T., Ueguchi, C., and Mizuno, T. 1995. Quantitative control of the stationary phase-specific sigma factor, σ^{S}, in *Escherichia coli*: Involvement of the nucleoid protein H-NS. EMBO J. 14:594–602.

141. Yildiz, F. H., and Schoolnik, G. K. 1998. Role of *rpoS* in stress survival and virulence of *Vibrio cholerae*. J. Bacteriol. 180(4):773–784.

142. Zhang, A., Altuvia, S., Tiwari, A., Argaman, L., Hengge-Aronis, R., and Storz, G. 1998. The OxyS regulatory RNA represses *rpoS* translation and binds the Hfq (HF-I) protein. EMBO J. **17**:6061–6068.
143. Zhou, Y., Gottesman, S., Hoskins, J. R., Maurizi, M. R., and Wickner, S. 2001. The RssB response regulator directly targets σ^S for degradation by ClpXP. Genes Dev. **15**:627–637.

CHAPTER 3

Surviving the immune response: an immunologist's perspective

David R. Katz and Gabriele Pollara

INTRODUCTION

The immune system exists to combat infection. By convention, there are considered to be two ways that the immune system combats infection. These are represented as the two poles of the horizontal axis in Fig. 3.1. The innate immune system – in effect the inflammatory response – is the non-specific component which responds indiscriminately to infection and other forms of injury and which uses phagocytosis as a major effector route. The adaptive immune system is specific and distinguishes not only between infection and injury but also between different types of infection. This system uses antibody and T-cell receptors for recognition. The fine specificity of the adaptive reaction, and how it is controlled, is one of the most remarkable of all biological processes and incorporates both memory – recall of past experience – and tolerance – recall to not respond.

From an immunologist's perspective, the ability of an infective agent to survive this double jeopardy immune response represents failure. But organisms can – and do – survive the response all the time. They do this by interfering with different stages in the natural history of the response, and six major examples of how this may happen are illustrated in Fig. 3.1 and Plate I. Figure 3.1 and Plate I also highlight the importance of location: fortunately most organisms are encountered at sites of immune surveillance; but, if they do happen to localise to a site of immune privilege, this may carry with it an intrinsic survival advantage.

Some of the features of these survival routes have been known for many years, based upon careful descriptive studies of bacterial infection. These studies have documented how variable the pattern of the bacterial approach to survival in the face of immunity may be. There appears to be a highly

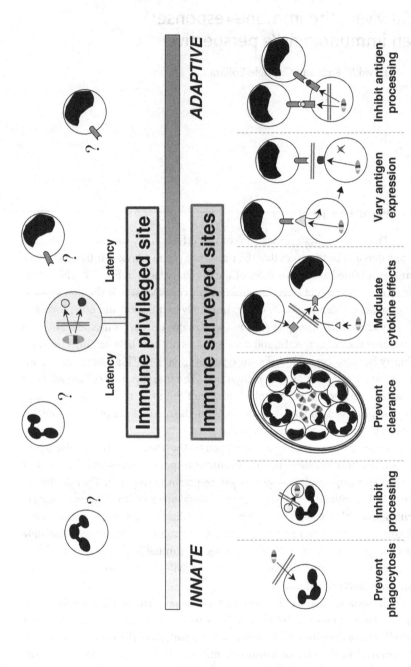

INNATE

Prevent phagocytosis

Inhibit processing

Prevent clearance

Modulate cytokine effects

Vary antigen expression

Inhibit antigen processing

ADAPTIVE

Immune surveyed sites

Immune privileged site

Latency

Latency

conserved relative "tropism" of bacteria for infectivity towards myeloid series (phagocytic) cells rather than epithelial cells or lymphocytes (125). Therefore the simplest route to promote survival in the face of immunity is for the bacterium to interfere with phagocytic pathways (reviewed in 31).

INTERFERENCE WITH PHAGOCYTOSIS

The presence of a capsule (such as is seen in the pneumococcus) can have this effect, despite the presence of an overall efficient innate response (71). Organisms may also produce mucoid secretions that interfere not only with barrier mechanisms (126) but also with activation of the alternate complement pathway, hence protecting themselves from lysis (12). *Neisseria spp* produce protease enzymes that cause mucosal IgA hydrolysis and thus survive in the presence of an opsonising adaptive reaction (80, 111). Once uptake has taken place there are still further opportunities for an organism to enhance survival, as they may produce a product that promotes sequestration within the phagocytic cell, as seen in some forms of staphylococcal infection (114). Similar mechanisms can also lead to survival of Gram-negative bacteria (16), and a combination of sequestration in phagocytic as well as non-phagocytic cells can provide a survival route for gastrointestinal pathogens such as Salmonella (161). Organisms may produce so-called "virulence factors," which are susceptible to neutrophil enzymes but which protect them in epithelial cells (59, 160); the corollary of this is that if the neutrophil enzyme pathway is itself inactivated for any reason, then the infection will be perpetuated (139).

LIMITATION OF CLEARANCE – THE GRANULOMA

There is one form of survival, by convention linked to the so-called delayed type hypersensitivity (type IV) reaction, which provides a unique setting for organisms to persist despite immune attack. This is the microenvironment of the granuloma. Leprosy is regarded as the paradigm, as it illustrates

Figure 3.1. Mechanisms of immune evasion. Pathogens utilise a range of mechanisms to subvert detection by the immune system across the spectra of innate and adaptive responses. *Upper panel*: Residing in immune privilege sites and low antigen production (e.g., herpes simplex virus latency in central nervous system) result in lack of detection by immune recognition mechanisms. *Lower panel – left to right*: (1) Prevent phagocytosis (e.g., *Streptococcus pneumoniae* polysaccharide). (2) Inhibit pathogen processing and destruction (e.g., *Mycobacterium tuberculosis* inhibition of phagolysosome fusion. (3) Prevent clearance (e.g., *Mycobacterium tuberculosis* granuloma induction). (4) Modulate cytokine effects (e.g., *Vaccinia* virus induced secretion of antagonistic cytokine homologue). (5) Vary antigen expression (e.g., *Trypansoma Cruzi* antigen switching). (6) Inhibit antigen processing and presentation (e.g., cytomegalovirus inhibition of MHC expression).

that both paucibacillary ("tuberculoid," because this is the more conventional appearance in *Mycobacterium tuberculosis* infection) and multibacillary ("lepromatous") symbiosis between macrophage and mycobacterium can occur (130). In parasitic diseases, too, the granuloma is important: even if by morphology the lesions are less well formed, as is seen in leishmaniasis, there is a similarity in the unusual relationship between cell and predator (113, 131, 143). The schistosome that survives is in fact the hallmark of the typical liver granulomatous lesion that is seen in bilharzia (137). The antigens that the remaining parasites themselves generate will in turn determine the pattern of host immune responsiveness, and hence their own further survival (133).

What are the general factors that control the granuloma and contribute to survival of an initiating factor, such as a bacterium or a parasite? The relative role of cell turnover must be important – these immune granulomas are sometimes described as "high turnover" where all stages of epithelioid differentiation are seen, there are growth factors and receptors present, and there is some local cell division (4, 145). There are several features of the infective agent that have been identified as being predisposing to both granuloma formation and survival of the initiating factor despite the ongoing response. The nature of the antigens exposed on the surface of the organism may have a selective role in determining the pattern of the response (53). The organism may be resistant to host enzymes (11). Uptake via complement receptors rather than by scavenger or Fc receptors will avoid the respiratory burst and thus lead to unexpected survival despite an innate immune reaction that is otherwise adequate (30). The organism may enter the cytoplasm directly rather than via a typical phagosome and thus avoid the potential trap of lysosome–phagosome fusion and processing (142). There may be interference in this phagosome–lysosome fusion stage (52). The requisite activation pathway necessary to limit survival and pathology may be inhibited (99). The capsule of the organism may mimic a foreign body and hence elicit a response resembling that seen to inert material (40). Once within the phagocyte the organism may shut down its own protein synthesis mechanism and thus remain dormant (66), and it also may interfere with the cytokine synthesis by the host cell (91, 127). As in other infections, there is a possibility that non-professional phagocytes – including epithelial cells and endothelial cells – may harbour the agent that promotes granuloma formation (20). Finally, the molecular mimicry theory has also been invoked and this too may be dependent upon sequestration and survival of an exogenous organism (33, 64).

Considerable interest in the pathogenesis of the granuloma in which mycobacteria can survive has been precipitated by the strong association of the diseases caused by these organisms with HIV-1 infection. In people with this condition there may be reactivation; and tuberculosis can take a "lepromatous" form with high organism counts. A T-cell role in determining this pattern has been suggested, and the possibility that there is activation of a specific "suppressor" form, which inhibits (or secretes a mediator inhibitor of) processing has been advocated (58). The proportional absence of CD4+ T cells in the HIV infection microenvironment may be a contributory factor (140, 162).

However, the question "What happens inside the macrophage to regulate mycobacterial survival?" has attracted most attention. This question is important not only in terms of mycobacterial diseases per se but also in conditions such as sarcoidosis and Crohn's disease, where an association has been suspected but no organisms and their remnants have been documented unequivocally (reviewed in (60) and (67)). Several studies have shown that the "mycobacterial vacuole" is part of the classic endosomal pathway, but it appears to be arrested at an early stage, in which cathepsin D is present and acidification is inefficient (13, 36, 149). A molecular approach has also recently been adopted to document how the type of infecting mycobacterium may contribute to the question of survival (158). There appear to be particular strains that cause infection but survive without causing "disease," and others that cause disease but are less readily transmissible. Furthermore, it has been suggested that the more virulent strains are in fact those that are more resistant to growth inhibition by the macrophages themselves (115).

GENETIC FACTORS

It is also in mycobacterial disease that the role of particular host genetic factors has been recognised as permissive for survival of infectious agents. Deletions in the interferon gamma gene lead to increased susceptibility to infection (19, 50). Mice with deletion of beta 2 microglobulin, which are unable to generate a T-cell response, do not show prolonged survival of the organisms, but they do show a different response pattern dependent upon gamma–delta T cells (88, 112). In man the absence of the interferon gamma receptor is associated with failure to eliminate atypical forms of mycobacteria (*M. avium intracellulare*), as well as *M. bovis* BCG (44, 141, 159). Deficiency of the interleukin 12 receptor (which leads to interferon gamma deficiency) also results in increased survival of the organisms and hence disease susceptibility (7, 47). Finally, there is increasing awareness that genetic polymorphisms may contribute to our understanding of susceptibility of mycobacterium to survive a response (62, 135). One of the best characterised of these is the natural resistance associated macrophage protein (NRAMP1) gene polymorphism, where particular alleles are linked to disease perpetuation despite the apparent adequacy of the ongoing response (17).

HELMINTH SURVIVAL OF THE IMMUNE RESPONSE

Of course, bacteria are not unique as invasive survivors of the immune response. Some of the most interesting examples are helminths, where survival, dependent on the nature of the immune response, can affect the ability to transfer disease. This is seen in Ascariasis, where in infected hosts the rise in IgE levels correlates inversely with the ability to excrete eggs. The lower the IgE, the more eggs excreted, and the worse the clinical disease. The consequence of this is that the lower the adaptive immune response, the greater the likelihood of transmission of the disease and hence survival of the helminth. Ascariasis is also of interest as an example in which a potent enzyme inhibitor (an anti-aspartic protease) is produced that will interfere with antigen processing pathways (74). Likewise, *Nippostrongylus braziliensis* has recently been shown to produce an enzyme with similar properties, suggesting that this may be a frequent helminth survival strategy (41).

ANTIGENIC VARIATION

It is perhaps not surprising that parasites are also particularly adept at surviving immune responses. They adapt in complex ways to their microenvironment, and from the point of view of survival this phenomenon can manifest as "antigenic variation." The classic example of this is seen in trypanosomiasis, where the parasite switches major variant surface glycoproteins and thus evades the effects of the immune response (reviewed in 14). The different forms of these proteins are controlled by upstream promoters that are active at different stages in the life cycle of the parasite (45, 54). One of the ligands for these proteins is mannose binding protein, which can opsonise the parasites, promote binding to macrophage receptors, and thus facilitate clearance by phagocytes (75). Hence changing your surface glycoproteins may allow you to evade mannose-binding-protein–mediated uptake and resultant elimination. In addition, trypanosomes can induce local CD8+ cells to produce gamma interferon but can continue to grow in the presence of the cytokine (42, 76). They also avoid complement-mediated lysis, probably as a result of synthesis of a surface protease (43).

In schistosomiasis there is a developmentally regulated protein, SmSP1, which has significant homology with human factor I light chain (37). It has been suggested that this may help to protect the parasite from the immune response. In addition, the parasite is capable of synthesising several serine proteases, including several different cathepsins (27, 28, 65, 103, 137, 138, 155). The granulomatous reaction to the ova appears to protect the host from the toxins produced by the parasite, but the precise roles of the T helper type 2 cytokines and of IgE – whether these are important in protecting host or parasite (81, 102, 123) – are unclear. In the presence of an ongoing granulomatous response there may also be an increased propensity for egg excretion – a further example of how delicate the balance between parasite survival and immunity can be.

Another parasite, *toxoplasma gondii*, characteristically renders the phago-cytic vacuoles fusion-incompetent (73, 95). In these vacuoles there is less acidification and parasite proteins accumulate. As they are isolated from the normal exocytic and secretory pathways, the vacuoles provide an excellent site for parasite long-term survival (109). It is interesting that in this example the role of antibodies is much clearer than in some of the other parasitic conditions, since opsonised parasites internalised via the Fc receptor pathway are eliminated (72, 110).

VIRAL SURVIVAL OF THE IMMUNE RESPONSE: LATENCY

In view of these multiple survival routes, various terms have been applied to describe how organisms do manage to cope with the onslaught of the immune system. One of the best known of these terms, "latency," is used by convention to describe the state in which the infective agent is present in some guise, with no active immune response to it, but where there is a clear risk of subsequent injury. It is in this context that viral survival is perhaps best considered.

The classic example of latency is where an initial viral infection resolves, and superficially it appears that the causative agent has been eliminated. However, this is followed some time later by a recrudescence. This may be in the form of a new version of the same response, as seen typically in herpes simplex virus infection (38, 63, 84, 121). There is a repeat of the innate–adaptive sequence, with remarkably little clinical influence of recall. This particular system has been considered to be a useful model to help us under-stand unexpected survival of a virus despite efficient and satisfactory adaptive immunity. Firstly the virus can bind to and travel up the axon, where (unlike at the periphery) it is relatively inaccessible to immune attack. Choosing your site of tropism may thus be important! Next, when it reaches the neuronal body, effectively only one viral gene is transcribed – the "latency associated transcript" (15, 34, 38, 146, 151, 153). Expression of this gene prevents death by apoptosis of the infected cell, thus perpetuating the presence of the virus in a live cell, but interfering at the same time with the cell's translational machinery (5, 119). If apoptosis were to occur, with resultant viral release, then this might cause injury, and an ongoing innate to adaptive immune response would ensue, which the virus might not survive. In fact, if any viral replication does occur locally, it is probably associated with a perfectly adequate response. It is, perhaps, relevant that CD8+ T cells have been shown to be necessary to maintain the latency state and that these provide a form of containment of the virus akin to the more familiar – and recognisable – granuloma outlined above (93). "Reactivation" – the process whereby the virus changes from latent to injury mode – requires the action of host neuronal proteins, probably cellular transcription factors, that interact with the viral genes which promote expression of the viral "immediate early" (IE) genes (51, 101, 152). Another route to reactivation (at least in experimental systems) results from deletion of the anti-apoptotic latency gene (68, 92). Irrespective of mechanism, both innate and adaptive components of this reactivation are usually seen at the periphery – the skin – rather than within the latency site itself. How this skin–nerve relationship works is not clear, which allows one to speculate and enunciate what may be an important principle: the virus survives the immune response not only because of its molecular properties but also because of its felicitous skill for localisation in an "immune privilege" site. Finally, these processes may well converge on what might be termed "molecular tropism" where neuronal specific transcription factors are what the virus needs to enter into (or exit from) latency (156).

In a second form of neurotropic herpesvirus infection, varicella–zoster, it has been known for many years at a clinical level that the survival pattern takes a different form from that seen with herpes simplex. Varicella–zoster infection leads to the disseminated cutaneous disease of multi-dermatome distribution that is known as chicken pox (147). Here there are both forms of adaptive response, antibody and cytotoxic T cell, and the local virus does not survive. However, the virus will survive elsewhere, again within neurones, and reactivate later in the clinical condition "shingles." Again this has a dermatome distribution, confirming the neural localisation. From the immunological point of view the two varicella–zoster associated diseases are very similar. The virus therefore avoids destruction at three stages: not only during the acute viraemic phase, but also during latency, and then during reactivation (3). It has been suggested that the way varicella–zoster does this is by synthesising an early gene, known as open reading frame (ORF) ORF66, which interferes with transport of major histocompatibility complex (MHC)

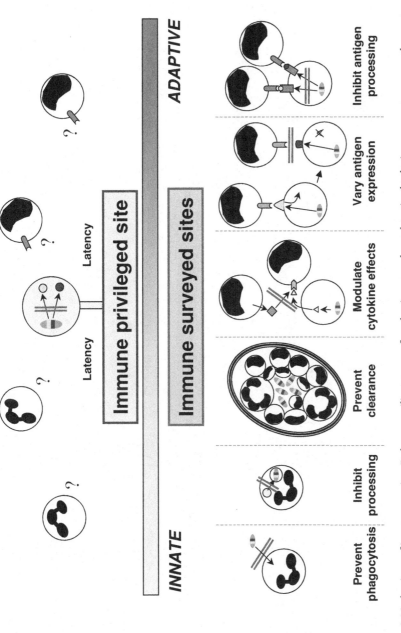

Plate I. Mechanisms of immune evasion. Pathogens utilise a range of mechanisms to subvert detection by the immune system across the spectra of innate and adaptive responses.

class I molecules from the Golgi apparatus to the cell surface (1, 2). The consequences of less class I MHC are that the cell is not recognised by CD8+ cytotoxic T cells during an initial critical phase in its reactivation, and therefore it survives long enough to cause the immunopathology (144). It is perhaps significant that the theme of interference runs through several examples of viral immune evasion and survival (see below).

A third herpesvirus, cytomegalovirus (CMV), is known for its capacity to develop lifelong latency, and this is seen within myeloid cells, i.e., within the immune system itself (96, 122) as well as in associated cells such as endothelium (56). Several different mechanisms have been implicated in the CMV survival pathway. The reactivation process for surviving virus is particularly important in immunodeficient people (22, 129) including transplant recipients and patients with iatrogenic immunosuppression (18, 136, 150) as well as in acquired immunodeficiency syndrome (77, 134). It is clear that inhibition of class II MHC expression in infected cells, mediated via the unique short (US) viral gene products, is not dependent upon viral replication and leads to failure to initiate CD4+ T-cell responses (154). These CMV US proteins also control class I MHC expression via a variety of different mechanisms, including inhibition of transport through TAP pores, reverse relocation of heavy chains into the cytosol where they can undergo proteosomal degradation, and inhibition of the calreticulin–peptide exchange mechanism within the endoplasmic reticulum (61).

The "newest" member of the herpesvirus family, HHV8 (KSHV), also downregulates class I MHC and takes the survival strategy one step further. The lack of class I makes the infected cells more susceptible to attack by natural killer (NK) cells. The virus survives by using the same K5 protein that is responsible for interference with class I to downregulate synthesis of the costimulatory molecules CD54 (ICAM-1) and CD86, both of which are required for NK activation. Hence, they successfully evade both killer pathways (69).

A different form of survival in the face of immune responsiveness is illustrated by some of the hepatitis viruses (35). Here the natural history of the hepatitis B virus is well known. Of course, the lack of an acute clinical episode does not mean that a potential blood donor is uninfected – so from their viewpoint these viruses have definitely survived the immune response. Both in the presence and absence of an acute phase, this virus causes "chronic injury" – in immunological terms, cytotoxic T cells are important in determining this (21, 78) but there is also longstanding production of abnormal local levels of cytokines and antibodies that target liver cells (57, 157). For example, local production of gamma interferon leads to an increase in expression of interferon receptors, as well as of class I and class II MHC, and there is local synthesis of tumour necrosis factor (TNF) and apoptosis (117). There is evidence that cytotoxic T cells directed against virally infected cells are present throughout this survival phase (98, 128), but it is not clear how they are controlled to ensure only partial elimination of infection. This is true both when the liver cells are infected with hepatitis B and in other forms of chronic liver disease, such as that caused by hepatitis C, where the acute phase is less well defined but where long-term viral survival is equally important (32). It has been suggested that secreted antigen may be a factor that promotes hepatitis B survival (104), and this may also apply for other hepatitis viruses.

SURVIVAL AND NEOPLASIA

Perhaps even more telling about the capacity for survival are the situations where the initial hepatitis virus exposure is occult, without any overt response, and where the first evidence of pathology takes the form of a virus-associated neoplasm (9, 87, 106). Initially inflammation and necrosis do not eliminate the virus but do initiate hepatocyte proliferation. Eventually there is viral integration and activation of hepatocyte growth regulatory genes, and the increased cell division increases the chances of a mutation (8). Because the host is only killed at a much later stage, the survival in the face of immunity is clearly important not only for the infected individual but also for transmission to society.

In the field of survival leading to neoplasia, the herpes viruses have also been implicated and studied in detail (82, 83). Epstein–Barr virus (EBV) infects both epithelial cells and B lymphocytes, and different ORFs help to determine tropism; and in B lymphocytes, the EBV "norm" is replication and transmission to daughter cells, rather than lysis (48, 55, 86). Two families of viral gene products, the EB nuclear antigens (EBNA) (90) and the latent membrane proteins (LMP) (108) have been shown to be actively transcribed in these cells and play a role in both cell protection and proliferation. For example, EBNA1 blocks processing of viral antigens in the proteosome (89) and thus inhibits killing of infected cells by cytotoxic T cells (24, 79); EBNA 3 is involved in control of costimulator molecule expression; and LMP1 protects the cells against apoptosis (70). Different patterns of relative EBNA and LMP expression have been linked with the infectious mononucleosis disease pattern, where there is a strong immune response, but the exact gene product responsible for the long-term post-disease viral survival in B memory lymphocytes remains controversial (23, 29, 46, 85). The intra-epithelial survival of EBV is also unusual, in that the most common type of tumour seen is a non-keratinising form of carcinoma, and there is a prominent admixture of lymphocytes, suggesting an ongoing immune reaction.

INTERFERENCE WITH MEDIATORS

Recent studies in several laboratories have suggested that there may be a common pathway for prolonging survival that has been adopted by several viruses. This is to interfere with the cytokine and chemokine networks. The concept of "sabotage" by this mechanism is attractive, as it is particularly likely to be short-range, acting at the interface between infected and potential responder cells. These studies have shown that some viruses encode homologues of pro-inflammatory mediators, such as interleukin 1 (IL-1) (26) and members of the TNF family (105), which can act antagonistically. Others encode for homologues of anti-inflammatory mediators (such as IL-10) (107). This is one way that HIV may act, as there is partial homology between the HIV-1 tat protein and monocyte chemotactic protein (MCP) (6, 10, 118). Other potential pathways illuminated by the experience with HIV-1 suggest that HIV-1 tat may induce TNF synthesis and hence increase local injury and cell death (100). Homology has also been described between viral proteins and cytokine binding proteins (25), and this has been invoked not only for chemokines – which can bind to the viral receptor rather than the cellular – but also in relationship to dummy binding of the pro-inflammatory mediator, IL-18 (116).

DICHOTOMY

Although most immunological interpretations of the term "surviving the immune response" focus inevitably on the way that causative organisms (or their components) fail to be eliminated, it is also possible to use the same term to describe an entirely different set of phenomena relating to the host. In these settings, of which type I anaphylaxis is the most striking example, initial exposure leads to a response. There is recall of that response pattern, so that when exposure recurs, a rapid reaction occurs leading to immune-mediated injury. There may be amplified immunogenicity, which can be life threatening, and thus one can use the term survival with some justification for the patients who do not succumb. Classically immune complex reactions can also cause serious systemic immune injury, as seen in serum sickness. Although these reactions are clearly different from those where the exogenous organism survives, nonetheless they are perhaps worthy of some consideration in this context.

Table 3.1. *Surviving the immune response:*
an immunologist's perspective

Avoid injury
Avoid uptake
Avoid processing
Avoid apoptosis
Avoid responder cell(s)

Select the right host
Select a protected host cell
Survive within the cell
Shift the response within the cell
Shift the nature of the response that is triggered

In both type I and some types of type IV reaction, the predominant triggers are either chemical or inert compounds rather than biologically active ones, and hence they can continue to act for extended periods. However, although the separation between different forms of immune injury and between biological and chemical triggers may be useful in general terms, there can be considerable overlap between these and the mechanisms of survival and resultant pathogenesis. For example, in the lung of the chronic asthmatic, the fungus *Aspergillus fumigatus* often survives and the resultant disease "allergic bronchopulmonary aspergillosis" has features of both type I and type III reactions (39, 94, 97). In some of the chemical irritant reactions a similar pattern has been described (49, 124, 132, 148). Essentially the chemical acts as a hapten and alters the immunogenicity of a protein; in the meantime the "parent" protein of course survives intact in the host.

Against this background it should be clear that, from an immunologist's viewpoint, surviving the immune response does not come as a surprise, given the variations on the theme available to organisms. These variations are summarised in Table 3.1.

Rather, perhaps, the surprise is that there are recall and memory responses to reactivation. Furthermore, immunologists would also argue that one has to be cautious about regarding survival as bad news. It may prove to be a very important factor in maintaining tolerance and memory. It is also possible that eliminating survival may lead to unexpected forms of immune injury when re-exposure occurs. Thus survival may prove to be a small price to pay for immune protection.

REFERENCES

1. Abendroth, A., and Arvin, A. M. 2001. Immune evasion as a pathogenic mechanism of varicella zoster virus. Semin. Immunol. **13**:27–39.
2. Abendroth, A., Lin, I., Slobedman, B., Ploegh, H., and Arvin, A. M. 2001. Varicella-zoster virus retains major histocompatibility complex class I proteins in the Golgi compartment of infected cells. J. Virol. **75**:4878–4888.
3. Abendroth, A., Morrow, G., Cunningham, A. L., and Slobedman, B. 2001. Varicella-zoster virus infection of human dendritic cells and transmission to T cells: Implications for virus dissemination in the host. J. Virol. **75**:6183–6192.
4. Adams, D. O., and Hamilton, T. A. 1989. The activated macrophage and granulomatous inflammation. Curr. Top. Pathol. **79**:151–167.
5. Ahmed, M., Lock, M., Miller, C. G., and Fraser, N. W. 2002. Regions of the herpes simplex virus type 1 latency-associated transcript that protect cells from apoptosis in vitro and protect neuronal cells in vivo. J. Virol. **76**:717–729.
6. Albini, A., Ferrini, S., Benelli, R., Sforzini, S., Giunciuglio, D., Aluigi, M. G., Proudfoot, A. E., Alouani, S., Wells, T. N., Mariani, G., Rabin, R. L., Farber, J. M., and Noonan, D. M. 1998. HIV-1 Tat protein mimicry of chemokines. Proc. Natl. Acad. Sci. USA **95**:13,153–13,158.
7. Altare, F., Durandy, A., Lammas, D., Emile, J. F., Lamhamedi, S., Le Deist, F., Drysdale, P., Jouanguy, E., Doffinger, R., Bernaudin, F., Jeppsson, O., Gollob, J. A., Meinl, E., Segal, A. W., Fischer, A., Kumararatne, D., and Casanova, J. L. 1998. Impairment of mycobacterial immunity in human interleukin-12 receptor deficiency. Science **280**:1432–1435.
8. Arbuthnot, P., Capovilla, A., and Kew, M. 2000. Putative role of hepatitis B virus X protein in hepatocarcinogenesis: Effects on apoptosis, DNA repair, mitogen-activated protein kinase and JAK/STAT pathways. J. Gastroenterol. Hepatol. **15**:357–368.
9. Arbuthnot, P., and Kew, M. 2001. Hepatitis B virus and hepatocellular carcinoma. Int. J. Exp. Pathol. **2**:77–100.
10. Arese, M., Ferrandi, C., Primo, L., Camussi, G., and Bussolino, F. 2001. HIV-1 Tat protein stimulates in vivo vascular permeability and lymphomononuclear cell recruitment. J. Immunol. **166**:1380–1388.
11. Azad, A. K., Sirakova, T. D., Fernandes, N. D., and Kolattukudy, P. E. 1997. Gene knockout reveals a novel gene cluster for the synthesis of a class of cell wall lipids unique to pathogenic mycobacteria. J. Biol. Chem. **272**:16,741–16,745.

87

SURVIVING THE IMMUNE RESPONSE

12. Baltimore, R. S., and Shedd, D. G. 1983. The role of complement in the opsonization of mucoid and non-mucoid strains of *Pseudomonas aeruginosa*. Pediatr. Res. **17**:952–958.

13. Barker, L. P., George, K. M., Falkow, S., and Small, P. L. 1997. Differential trafficking of live and dead *Mycobacterium marinum* organisms in macrophages. Infect. Immun. **65**:1497–1504.

14. Barry, J. D., and McCulloch, R. 2001. Antigenic variation in trypanosomes: Enhanced phenotypic variation in a eukaryotic parasite. Adv. Parasitol. **49**:1–70.

15. Batchelor, A. H., and O'Hare, P. 1990. Regulation and cell-type-specific activity of a promoter located upstream of the latency-associated transcript of herpes simplex virus type 1 required for cell-type-specific activity. J. Virol. **64**:3269–3279.

16. Behlau, I., and Miller, S. I. 1993. A PhoP-repressed gene promotes *Salmonella typhimurium* invasion of epithelial cells. J. Bacteriol. **175**:4475–4484.

17. Bellamy, R., Ruwende, C., Corrah, T., McAdam, K. P., Whittle, H. C., and Hill, A. V. 1998. Variations in the NRAMP1 gene and susceptibility to tuberculosis in West Africans. N. Engl. J. Med. **338**:640–644.

18. Belles-Isles, M., Houde, I., Lachance, J. G., Noel, R., Kingma, I., and Roy, R. 1998. Monitoring of cytomegalovirus infections by the CD8+CD38+ T-cell subset in kidney transplant recipients. Transplantation **65**:279–282.

19. Benini, J., Ehlers, E. M., and Ehlers, S. 1999. Different types of pulmonary granuloma necrosis in immunocompetent vs. TNFRp55-gene-deficient mice aerogenically infected with highly virulent *Mycobacterium avium*. J. Pathol. **189**:127–137.

20. Bermudez, L. E., Sangari, F. J., Kolonoski, P., Petrofsky, M., and Goodman, J. 2002. The efficiency of the translocation of *Mycobacterium tuberculosis* across a bilayer of epithelial and endothelial cells as a model of the alveolar wall is a consequence of transport within mononuclear phagocytes and invasion of alveolar epithelial cells. Infect. Immun. **70**:140–146.

21. Bertoletti, A., Costanzo, A., Chisari, F. V., Levrero, M., Artini, M., Sette, A., Penna, A., Giuberti, T., Fiaccadori, F., and Ferrari, C. 1994. Cytotoxic T lymphocyte response to a wild type hepatitis B virus epitope in patients chronically infected by variant viruses carrying substitutions within the epitope. J. Exp. Med. **180**:933–943.

22. Betts, R. F., and Hanshaw, J. B. 1977. Cytomegalovirus (CMV) in the compromised host(s). Annu. Rev. Med. **28**:103–110.

23. Bishop, G. A., and Hostager, B. S. 2001. Signaling by CD40 and its mimics in B cell activation. Immunol. Res. **24**:97–109.

24. Blake, N., Haigh, T., Shaka'a, G., Croom-Carter, D., and Rickinson, A. 2000. The importance of exogenous antigen in priming the human CD8+ T cell response: Lessons from the EBV nuclear antigen EBNA1. J. Immunol. **165**:7078–7087.

25. Bodaghi, B., Jones, T. R., Zipeto, D., Vita, C., Sun, L., Laurent, L., Arenzana-Seisdedos, F., Virelizier, J. L., and Michelson, S. 1998. Chemokine sequestration by viral chemoreceptors as a novel viral escape strategy: Withdrawal of chemokines from the environment of cytomegalovirus-infected cells. J. Exp. Med. **188**:855–866.

26. Bowie, A., Kiss-Toth, E., Symons, J. A,. Smith, G. L., Dower, S. K., and O'Neill, L. A. 2000. A46R and A52R from vaccinia virus are antagonists of host IL-1 and toll-like receptor signaling. Proc. Natl. Acad. Sci. USA **97**: 10,162–10,167.

27. Brindley, P. J., Kalinna, B. H., Wong, J. Y., Bogitsh, B. J., King, L. T., Smyth, D. J., Verity, C. K., Abbenante, G., Brinkworth, R. I., Fairlie, D. P., Smythe, M. L., Milburn, P. J., Bielefeldt-Ohmann, H., Zheng, Y., and McManus, D. P. 2001. Proteolysis of human hemoglobin by schistosome cathepsin D. Mol. Biochem. Parasitol. **112**:103–112.

28. Caffrey, C. R., and Ruppel, A. 1997. Cathepsin B-like activity predominates over cathepsin L-like activity in adult *Schistosoma mansoni* and *S. japonicum*. Parasitol. Res. **83**:632–635.

29. Cahir McFarland, E. D., Izumi, K. M., and Mosialos, G. 1999. Epstein-Barr virus transformation: Involvement of latent membrane protein 1-mediated activation of NF-kappaB. Oncogene **18**:6959–6964.

30. Caron, E., and Hall, A. 1998. Identification of two distinct mechanisms of phagocytosis controlled by different Rho GTPases. Science **282**:1717–1721.

31. Celli, J., and Finlay, B. B. 2002. Bacterial avoidance of phagocytosis. Trends Microbiol. **10**:232–237.

32. Cerny, A., and Chisari, F. V. 1999. Pathogenesis of chronic hepatitis C: Immunological features of hepatic injury and viral persistence. Hepatology **30**:595–601.

33. Chan, E., Fossati, G., Giuliani, P., Lucietto, P., Zaliani, A., Coates, A. R., and Mascagni, P. 1995. Sequence and structural homologies between *M. tuberculosis* chaperonin 10 and the MHC class I/II peptide binding cleft. Biochem. Biophys. Res. Commun. **211**:14–20.

34. Chen, X., Schmidt, M. C., Goins, W. F., and Glorioso, J. C. 1905. Two herpes simplex virus type 1 latency-active promoters differ in their contributions to latency-associated transcript expression during lytic and latent infections. J. Virol. **69**:7899–7908.

35. Chisari, F. V., and Ferrari, C. 1995. Hepatitis B virus immunopathogenesis. Annu. Rev. Immunol. **13**:29–60.

36. Clemens, D. L., and Horwitz, M. A. 1995. Characterization of the *Mycobacterium tuberculosis* phagosome and evidence that phagosomal maturation is inhibited. J. Exp. Med. **181**:257–270.

37. Cocude, C., Pierrot, C., Cetre, C., Fontaine, J., Godin, C., Capron, A., and Khalife, J. 1999. Identification of a developmentally regulated *Schistosoma mansoni* serine protease homologous to mouse plasma kallikrein and human factor I. Parasitology **118**:389–396.

38. Cohrs, R. J., and Gilden, D. H. 2001. Human herpesvirus latency. Brain Pathol. **11**:465–474.

39. Cromwell, O., Moqbel, R., Fitzharris, P., Kurlak, L., Harvey, C., Walsh, G. M., Shaw, R. J., and Kay, A. B. 1988. Leukotriene C4 generation from human eosinophils stimulated with IgG-*Aspergillus fumigatus* antigen immune complexes. J. Allergy Clin. Immunol. **82**:535–543.

40. Daffe, M., and Etienne, G. 1999. The capsule of *Mycobacterium tuberculosis* and its implications for pathogenicity. Tuber. Lung Dis. **79**:153–169.

41. Dainichi, T., Maekawa, Y., Ishii, K., Zhang, T., Nashed, B. F., Sakai, T., Takashima, M., and Himeno, K. 2001. Nippocystatin, a cysteine protease inhibitor from *Nippostrongylus brasiliensis*, inhibits antigen processing and modulates antigen-specific immune response. Infect. Immun. **69**:7380–7386.

42. Darji, A., Beschin, A., Sileghem, M., Heremans, H., Brys, L., and De Baetselier, P. 1996. In vitro simulation of immunosuppression caused by Trypanosoma brucei: Active involvement of gamma interferon and tumor necrosis factor in the pathway of suppression. Infect. Immun. **64**:1937–1943.

43. Donelson, J. E., Hill, K. L., and El-Sayed, N. M. 1998. Multiple mechanisms of immune evasion by African trypanosomes. Mol. Biochem. Parasitol. **91**:51–66.

44. Dorman, S. E., and Holland, S. M. 1998. Mutation in the signal-transducing chain of the interferon-gamma receptor and susceptibility to mycobacterial infection. J. Clin. Invest. **101**:2364–2369.

45. Downey, N., and Donelson, J. E. 1999. Search for promoters for the GARP and rRNA genes of *Trypanosoma congolense*. Mol. Biochem. Parasitol. **104**:25–38.

46. D'Souza, B., Rowe, M., and Walls, D. 2000. The bfl-1 gene is transcriptionally upregulated by the Epstein-Barr virus LMP1, and its expression promotes the survival of a Burkitt's lymphoma cell line. J. Virol. **74**:6652–6658.

47. Emile, J. F., Patey, N., Altare, F., Lamhamedi, S., Jouanguy, E., Boman, F., Quillard, J., Lecomte-Houcke, M., Verola, O., Mousnier, J. F., Dijoud, F.,

Blanche, S., Fischer, A., Brousse, N., and Casanova, J. L. 1997. Correlation of granuloma structure with clinical outcome defines two types of idiopathic disseminated BCG infection. J. Pathol. **181**:25–30.

48. Faulkner, G. C., Krajewski, A. S., and Crawford, D. H. 2000. The ins and outs of EBV infection. Trends Microbiol. **8**:185–189.

49. Feldmann, M. 1972. Induction of immunity and tolerance in vitro by hapten protein conjugates. I. The relationship between the degree of hapten conjugation and the immunogenicity of dinitrophenylated polymerized flagellin. J. Exp. Med. **135**:735–753.

50. Flesch, I. E., Hess, J. H., Huang, S., Aguet, M., Rothe, J., Bluethmann, H., and Kaufmann, S. H. 1995. Early interleukin 12 production by macrophages in response to mycobacterial infection depends on interferon gamma and tumor necrosis factor alpha. J. Exp. Med. **181**:1615–1621.

51. Franchini, M., Abril, C., Schwerdel, C., Ruedl, C., Ackermann, M., and Suter, M. 2001. Protective T-cell-based immunity induced in neonatal mice by a single replicative cycle of herpes simplex virus. J. Virol. **75**:83–89.

52. Fratti, R. A., Backer, J. M., Gruenberg, J., Corvera, S., and Deretic, V. 2001. Role of phosphatidylinositol 3-kinase and Rab5 effectors in phagosomal biogenesis and mycobacterial phagosome maturation arrest. J. Cell Biol. **154**:631–644.

53. Gilleron, M., Ronet, C., Mempel, M., Monsarrat, B., Gachelin, G., and Puzo, G. 2001. Acylation state of the phosphatidylinositol mannosides from Mycobacterium bovis bacillus Calmette Guerin and ability to induce granuloma and recruit natural killer T cells. J. Biol. Chem. **276**:34,896–34,904.

54. Graham, S. V., Wymer, B., and Barry, J. D. 1998. A trypanosome metacyclic VSG gene promoter with two functionally distinct, life cycle stage-specific activities. Nucl. Acids Res. **26**:1985–1990.

55. Griffin, B. E., and Xue, S. A. 1998. Epstein-Barr virus infections and their association with human malignancies: Some key questions. Ann. Med. **30**:249–259.

56. Guetta, E., Scarpati, E. M., and DiCorleto, P. E. 2001. Effect of cytomegalovirus immediate early gene products on endothelial cell gene activity. Cardiovasc. Res. **50**:538–546.

57. Guidotti, L. G., and Chisari, F. V. 1996. To kill or to cure: Options in host defense against viral infection. Curr. Opin. Immunol. **8**:478–483.

58. Haanen, J. B., Ottenhoff, T. H., Lai, A., Fat, R. F., Soebono, H., Spits, H., and de Vries, R. R. 1990. *Mycobacterium leprae* – specific T cells from a tuberculoid leprosy patient suppress HLA-DR3–restricted T cell responses to an immunodominant epitope on 65-kDa hsp of mycobacteria. J. Immunol. **145**: 3898–3904.

59. Hale, T. H. 1991. Genetic basis of virulence in Shigella species. Microbiol. Rev. **55**:206–224.

60. Hance, A. J. 1998. The role of mycobacteria in the pathogenesis of sarcoidosis. Semin. Resp. Infect. **13**:197–205.

61. Hengel, H., Koopmann, J. O., Flohr, T., Muranyi, W., Goulmy, E., Hammerling, G. J., Koszinowski, U. H., and Momburg, F. 1997. A viral ER-resident glycoprotein inactivates the MHC-encoded peptide transporter. Immunity **6**:623–632.

62. Hill, A. V. 1998. The immunogenetics of human infectious diseases. Annu. Rev. Immunol. **16**:593–617.

63. Hill, T. J., Field, H. J., and Roome, A. P. 1972. Intra-axonal location of herpes simplex virus particles. J. Gen. Virol. **15**:233–235.

64. Hogervorst, E. J., Boog, C. J., Wagenaar, J. P., Wauben, M. H., Van der Zee, R., and Van Eden, W. 1991. T cell reactivity to an epitope of the mycobacterial 65-kDa heat-shock protein (hsp 65) corresponds with arthritis susceptibility in rats and is regulated by hsp 65-specific cellular responses. Eur. J. Immunol. **21**:1289–1296.

65. Hola-Jamriska, L., King, L. T., Dalton, J. P., Mann, V. H., Aaskov, J. G., and Brindley, P. J. 2000. Functional expression of dipeptidyl peptidase I (Cathepsin C) of the oriental blood fluke *Schistosoma japonicum* in Trichoplusia ni insect cells. Protein Expr. Purif. **19**:384–392.

66. Hu, Y. M., Butcher, P. D., Sole, K., Mitchison, D. A., and Coates, A. R. 1998. Protein synthesis is shutdown in dormant *Mycobacterium tuberculosis* and is reversed by oxygen or heat shock. FEMS Microbiol Lett. **158**:139–145.

67. Hubbard, J., and Surawicz, C. M. 1999. Etiological role of mycobacterium in Crohn's disease: An assessment of the literature. Dig. Dis. **17**:6–13.

68. Inman, M., Perng, G. C., Henderson, G., Ghiasi, H., Nesburn, A. B., Wechsler, S. L., and Jones, C. 2001. Region of herpes simplex virus type 1 latency-associated transcript sufficient for wild-type spontaneous reactivation promotes cell survival in tissue culture. J. Virol. **75**:3636–3646.

69. Ishido, S., Choi, J. K., Lee, B. S., Wang, C., DeMaria, M., Johnson, R. P., Cohen, G. B., and Jung, J. U. 2000. Inhibition of natural killer cell-mediated cytotoxicity by Kaposi's sarcoma-associated herpesvirus K5 protein. Immunity **13**:365–374.

70. Izumi, K. M., Cahir McFarland, E. D., Ting, A. T., Riley, E. A., Seed, B., and Kieff, E. D. 1999. The Epstein-Barr virus oncoprotein latent membrane protein 1 engages the tumor necrosis factor receptor-associated proteins TRADD and receptor-interacting protein (RIP) but does not induce apoptosis or require RIP for NF-kappaB activation. Mol. Cell Biol. **19**:5759–5767.

KATZ AND POLLARA

71. Jarva, H., Janulczyk, R., Hellwage, J., Zipfel, P. F., Bjorck, L., and Meri, S. 2002. *Streptococcus pneumoniae* evades complement attack and opsonophagocytosis by expressing the pspC locus-encoded Hic protein that binds to short consensus repeats 8–11 of factor H. J. Immunol. **168**:1886–1894.

72. Joiner, K. A. 1993. Cell entry by *Toxoplasma gondii*: All paths do not lead to success. Res. Immunol. **144**:34–38.

73. Jones, T. C., Yeh, S., and Hirsch, J. G. 1972. The interaction between *Toxoplasma gondii* and mammalian cells. I. Mechanism of entry and intracellular fate of the parasite. J. Exp. Med. **136**:1157–1172.

74. Kageyama, T. 1998. Molecular cloning, expression and characterization of an Ascaris inhibitor for pepsin and cathepsin E. Eur. J. Biochem. **253**:804–809.

75. Kahn, S. J., Wleklinski, M., Ezekowitz, R. A., Coder, D., Aruffo, A., and Farr, A. 1996. The major surface glycoprotein of Trypanosoma cruzi amastigotes are ligands of the human serum mannose-binding protein. Infect. Immun. **64**:2649–2656.

76. Kahn, S. J., and Wleklinski, M. 1997. The surface glycoproteins of *Trypanosoma cruzi* encode a superfamily of variant T cell epitopes. J. Immunol. **159**:4444–4451.

77. Kaur, A., Rosenzweig, M., and Johnson, R. P. 2000. Immunological memory and acquired immunodeficiency syndrome pathogenesis. Philos. Trans. R. Soc. Lond. B Biol. Sci. **355**:381–390.

78. Khakoo, S. I., Ling, R., Scott, I., Dodi, A. I., Harrison, T. J., Dusheiko, G. M., and Madrigal, J. A. 2000. Cytotoxic T lymphocyte responses and CTL epitope escape mutation in HBsAg, anti-HBe positive individuals. Gut 47:137–143.

79. Khanna, R., Moss, D. J., and Burrows, S. R. 1999. Vaccine strategies against Epstein-Barr virus-associated diseases: Lessons from studies on cytotoxic T-cell-mediated immune regulation. Immunol. Rev. **170**:49–64.

80. Kilian, M., Reinholdt, J., Lomholt, H., Poulsen, K., and Frandsen, E. V. 1996. Biological significance of IgA1 proteases in bacterial colonization and pathogenesis: Critical evaluation of experimental evidence. APMIS **104**:321–338.

81. King, C. L., Malhotra, I., Mungai, P., Wamachi, A., Kioko, J., Muchiri, E., and Ouma, J. H. 2001. *Schistosoma haematobium*-induced urinary tract morbidity correlates with increased tumor necrosis factor-alpha and diminished interleukin-10 production. J. Infect. Dis. **184**:1176–1182.

82. Klein, E. 1998. The complexity of the Epstein-Barr virus infection in humans. Pathol. Oncol. Res. **4**:3–7.

83. Klein, G. 2002. Perspectives in studies of human tumor viruses. Front Biosci. **7**:d268–274.

93

84. Klein, R. J. 1982. The pathogenesis of acute, latent and recurrent herpes simplex virus infections. Arch. Virol. **72**:143–168.

85. Knecht, H., Berger, C., McQuain, C., Rothenberger, S., Bachmann, E., Martin, J., Esslinger, C., Drexler, H. G., Cai, Y. C., Quesenberry, P. J., and Odermatt, B. F. 1999. Latent membrane protein 1 associated signaling pathways are important in tumor cells of Epstein-Barr virus negative Hodgkin's disease. Oncogene **18**:7161–7167.

86. Knecht, H., Berger, C., Rothenberger, S., Odermatt, B. F., and Brousset, P. 2001. The role of Epstein-Barr virus in neoplastic transformation. Oncology **60**:289–302.

87. Koike, K., Tsutsumi, T., Fujie, H., Shintani, Y., and Kyoji, M. 2002. Molecular mechanism of viral hepatocarcinogenesis. Oncology 62 (Suppl 1):29–37.

88. Ladel, C. H., Daugelat, S., and Kaufmann, S. H. 1995. Immune response to Mycobacterium bovis bacille Calmette Guerin infection in major histocompatibility complex class I- and II-deficient knock-out mice: Contribution of CD4 and CD8 T cells to acquired resistance. Eur. J. Immunol. **25**:377–384.

89. Levitskaya, J., Sharipo, A., Leonchiks, A., Ciechanover, A., and Masucci, M. G. 1997. Inhibition of ubiquitin/proteasome-dependent protein degradation by the Gly-Ala repeat domain of the Epstein-Barr virus nuclear antigen 1. Proc. Natl. Acad. Sci. USA **94**:12,616–12,621.

90. Levitsky, V., Zhang, Q. J., Levitskaya, J., Kurilla, M. G., and Masucci, M. G. 1997. Natural variants of the immunodominant HLA A11-restricted CTL epitope of the EBV nuclear antigen-4 are nonimmunogenic due to intracellular dissociation from MHC class I:peptide complexes. J. Immunol. **159**:5383–5390.

91. Lewthwaite, J. C., Coates, A. R., Tormay, P., Singh, M., Mascagni, P., Poole, S., Roberts, M., Sharp, L., and Henderson, B. 2001. *Mycobacterium tuberculosis* chaperonin 60.1 is a more potent cytokine stimulator than chaperonin 60.2 (Hsp 65) and contains a CD14-binding domain. Infect. Immun. **69**:7349–7355.

92. Liacono, C. M., Myers, R., and Mitchell, W. J. 2002. Neurons differentially activate the herpes simplex virus type 1 immediate-early gene ICP0 and ICP27 promoters in transgenic mice. J. Virol. **76**:2449–2459.

93. Liu, T., Khanna, K. M., Chen, X., Fink, D. J., and Hendricks, R. L. 2000. CD8(+) T cells can block herpes simplex virus type 1 (HSV-1) reactivation from latency in sensory neurons. J. Exp. Med. **191**:1459–1466.

94. Longbottom, J. L. 1983. Allergic bronchopulmonary aspergillosis: Reactivity of IgE and IgG antibodies with antigenic components of *Aspergillus fumigatus* (IgE/IgG antigen complexes). J. Allergy Clin. Immunol. **72**:668–675.

95. Lycke, E., Carlberg, K., and Norrby, R. 1975. Interactions between *Toxoplasma gondii* and its host cells: Function of the penetration-enhancing factor of toxoplasma. Infect. Immun. 11:853–861.

96. Maciejewski, J. P., and St. Jeor, S. C. 1999. Human cytomegalovirus infection of human hematopoietic progenitor cells. Leuk. Lymphoma 33:1–13.

97. Madan, T., Banerjee, B., Bhatnagar, P. K., Shah, A., and Sarma, P. U. 1997. Identification of 45 kD antigen in immune complexes of patients of allergic bronchopulmonary aspergillosis. Mol. Cell Biochem. 166:111–116.

98. Maini, M. K., Boni, C., Lee, C. K., Larrubia, J. R., Reignat, S., Ogg, G. S., King, A. S., Herberg, J., Gilson, R., Alisa, A., Williams, R., Vergani, D., Naoumov, N. V., Ferrari, C., and Bertoletti, A. 2000. The role of virus-specific CD8(+) cells in liver damage and viral control during persistent hepatitis B virus infection. J. Exp. Med. 191:1269–1280.

99. Manca, C., Paul, S., Barry, C. E., III, Freedman, V. H., and Kaplan G. 1999. *Mycobacterium tuberculosis* catalase and peroxidase activities and resistance to oxidative killing in human monocytes in vitro. Infect. Immun. 67:74–79.

100. Mayne, M., Bratanich, A. C., Chen, P., Rana, F., Nath, A., and Power, C. 1998. HIV-1 tat molecular diversity and induction of TNF-alpha: Implications for HIV-induced neurological disease. Neuroimmunomodulation 5:184–192.

101. McKenna, D. B., Neill, W. A., and Norval, M. 2001. Herpes simplex virus-specific immune responses in subjects with frequent and infrequent orofacial recrudescences. Br. J. Dermatol. 144:459–464.

102. McKerrow, J. H. 1997. Cytokine induction and exploitation in schistosome infections. Parasitology 115 (Suppl):S107–112.

103. McKerrow, J. H. 1999. Development of cysteine protease inhibitors as chemotherapy for parasitic diseases: Insights on safety, target validation, and mechanism of action. Int. J. Parasitol. 29:833–837.

104. Milich, D. R., Chen, M. K., Hughes, J. L., and Jones, J. E. 1998. The secreted hepatitis B precore antigen can modulate the immune response to the nucleocapsid: A mechanism for persistence. J. Immunol. 160:2013–2021.

105. Misawa, K., Nosaka, T., Kojima, T., Hirai, M., and Kitamura, T. 2000. Molecular cloning and characterization of a mouse homolog of human TNFSF14, a member of the TNF superfamily. Cytogenet. Cell Genet. 89:89–91.

106. Monto, A., and Wright, T. L. 2001. The epidemiology and prevention of hepatocellular carcinoma. Semin. Oncol. 28:441–449.

107. Moore, K. W., Vieira, P., Fiorentino, D. F., Trounstine, M. L., Khan, T. A., and Mosmann, T. R. 1990. Homology of cytokine synthesis inhibitory factor (IL-10) to the Epstein-Barr virus gene BCRFI. Science 248:1230–1234.

108. Moorthy, R. K., and Thorley-Lawson, D. A. 1993. Biochemical, genetic, and functional analyses of the phosphorylation sites on the Epstein-Barr

virus-encoded oncogenic latent membrane protein LMP-1. J. Virol. **67**:2637–2645.

109. Mordue, D. G., Hakansson, S., Niesman, I., and Sibley, L. D. 1999. *Toxoplasma gondii* resides in a vacuole that avoids fusion with host cell endocytic and exocytic vesicular trafficking pathways. Exp. Parasitol. **92**:87–99.

110. Mordue, D. G., and Sibley, L. D. 1997. Intracellular fate of vacuoles containing *Toxoplasma gondii* is determined at the time of formation and depends on the mechanism of entry. J. Immunol. **159**:4452–4459.

111. Mulks, M. H., and Plaut, A. G. 1978. IgA protease production as a characteristic distinguishing pathogenic from harmless neisseriaceae. N. Engl. J. Med. **299**:973–976.

112. Muller, D., Pakpreo, P., Filla, J., Pederson, K., Cigel, F., and Malkovska, V. 1995. Increased gamma-delta T-lymphocyte response to *Mycobacterium bovis* BCG in major histocompatibility complex class I-deficient mice. Infect. Immun. **63**:2361–2366.

113. Murray, H. W. 2001. Tissue granuloma structure-function in experimental visceral leishmaniasis. Int. J. Exp. Pathol. **82**:249–267.

114. Nilsson, I. M., Lee, J. C., Bremell, T., Ryden, C., and Tarkowski, A. 1997. The role of staphylococcal polysaccharide microcapsule expression in septicaemia and septic arthritis. Infect Immun. **65**:4216–4221.

115. North, R. J., and Izzo, A. A. 1993. Mycobacterial virulence. Virulent strains of Mycobacteria tuberculosis have faster in vivo doubling times and are better equipped to resist growth-inhibiting functions of macrophages in the presence and absence of specific immunity. J. Exp. Med. **177**:1723–1733.

116. Novick, D., Kim, S. H., Fantuzzi, G., Reznikov, L. L., Dinarello, C. A., and Rubinstein, M. 1999. Interleukin-18 binding protein: A novel modulator of the Th1 cytokine response. Immunity **10**:127–136.

117. Ohta, A., Sekimoto, M., Sato, M., Koda, T., Nishimura, S., Iwakura, Y., Sekikawa, K., and Nishimura, T. 2002. Indispensable role for TNF-alpha and IFN-gamma at the effector phase of liver injury mediated by Th1 cells specific to hepatitis B virus surface antigen. J. Immunol. **165**:956–961.

118. Park, I. W., Wang, J. F., and Groopman, J. E. 2001. HIV-1 Tat promotes monocyte chemoattractant protein-1 secretion followed by transmigration of monocytes. Blood **97**:352–358.

119. Perng, G. C., Jones, C., Ciacci-Zanella, J., Stone, M., Henderson, G., Yukht, A., Slanina, S. M., Hofman, F. M., Ghiasi, H., Nesburn, A. B., and Wechsler, S. L. 2000. Virus-induced neuronal apoptosis blocked by the herpes simplex virus latency-associated transcript. Science **287**:1500–1503.

120. Pieters, J., and Gatfield, J. 2002. Hijacking the host: Survival of pathogenic mycobacteria inside macrophages. Trends Microbiol. **10**:142–146.

KATZ AND POLLARA

121. Price, R. W., Katz, B. J., and Notkins, A. L. 1975. Latent infection of the peripheral ANS with herpes simplex virus. Nature **257**:686–688.

122. Prosch, S., Docke, W. D., Reinke, P., Volk, H. D., and Kruger, D. H. 1999. Human cytomegalovirus reactivation in bone-marrow-derived granulocyte/monocyte progenitor cells and mature monocytes. Intervirology **42**:308–313.

123. Qadir, K., Metwali, A., Blum, A. M., Li, J., Elliott, D. E., and Weinstock, J. V. 2001. TGF-beta and IL-10 regulation of IFN-gamma produced in Th2–type schistosome granulomas requires IL-12. Am. J. Physiol. Gastrointest. Liver Physiol. **281**:G940–946.

124. Rajewsky, K., Schirrmacher, V., Nase, S., and Jerne, N. K. 1969. The requirement of more than one antigenic determinant for immunogenicity. J. Exp. Med. **129**:1131–1143.

125. Ramet, M., Manfruelli, P., Pearson, A., Mathey-Prevot, B., and Ezekowitz, R. A. 2002. Functional genomic analysis of phagocytosis and identification of a Drosophila receptor for E. coli. Nature **416**:644–648.

126. Ramphal, R., and Vishwanath, S. 1987. Why is Pseudomonas the colonizer and why does it persist? Infection **15**:281–287.

127. Redpath, S., Ghazal, P., and Gascoigne, N. R. 2001. Hijacking and exploitation of IL-10 by intracellular pathogens. Trends Microbiol. **9**:86–92.

128. Rehermann, B., Ferrari, C., Pasquinelli, C., and Chisari, F. V. 1996. The hepatitis B virus persists for decades after patients' recovery from acute viral hepatitis despite active maintenance of a cytotoxic T-lymphocyte response. Nat. Med. **2**:1104–1108.

129. Riddell, S. R. 1995. Pathogenesis of cytomegalovirus pneumonia in immunocompromised hosts. Semin. Respir. Infect. **10**:199–208.

130. Ridley, D. S. 1974. Histological classification and the immunological spectrum of leprosy. Bull. World Health Organ. **51**:451–465.

131. Ridley, M. J., and Ridley, D. S. 1986. Monocyte recruitment, antigen degradation and localization in cutaneous leishmaniasis. Br. J. Exp. Pathol. **67**:209–218.

132. Rittenberg, M. B., and Amkraut, A. A. 1966. Immunogenicity of trinitrophenyl-hemocyanin: Production of primary and secondary anti-hapten precipitins. J. Immunol. **97**:421–430.

133. Ross, A. G., Bartley, P. B., Sleigh, A. C., Olds, G. R., Li, Y., Williams, G. M., and McManus, D. P. 2002. Schistosomiasis. N. Engl. J. Med. **346**: 1212–1220.

134. Rowbottom, A. W., Lepper, M. W., Sharpstone, D., and Gazzard, B. 1998. Cytomegalovirus (CMV) infection in AIDS patients is associated with a CD3 receptor-mediated T cell hyporesponsiveness. Clin. Exp. Immunol. **111**:559–563.

135. Roy, S., Frodsham, A., Saha, B., Hazra, S. K., Mascie-Taylor, C. G., and Hill, A. V. 1999. Association of vitamin D receptor genotype with leprosy type. J. Infect. Dis. **179**:187–191.

136. Rubin, R. H. 1990. Impact of cytomegalovirus infection on organ transplant recipients. Rev. Infect. Dis. 12 (Suppl 7):S754–766.

137. Rutitzky, L. I., Hernandez, H. J., and Stadecker, M. J. 2001. Th1–polarizing immunization with egg antigens correlates with severe exacerbation of immunopathology and death in schistosome infection. Proc. Natl. Acad. Sci. USA **98**:13,243–13,248.

138. Sajid, M., and McKerrow, J. H. 2002. Cysteine proteases of parasitic organisms. Mol. Biochem. Parasitol. **120**:1–21.

139. Sallenave, J. M. 2000. The role of secretory leukocyte proteinase inhibitor and elafin (elastase-specific inhibitor/skin-derived antileukoprotease) as alarm antiproteinases in inflammatory lung disease. Respir. Res. 1:87–92.

140. Sampaio, E. P., Caneshi, J. R., Nery, J. A., Duppre, N. C., Pereira, G. M., Vieira, L. M., Moreira, A. L., Kaplan, G., and Sarno, E. N. 1995. Cellular immune response to *Mycobacterium leprae* infection in human immunodeficiency virus-infected individuals. Infect. Immun. **63**:1848–1854.

141. Sasaki, Y., Nomura, A., Kusuhara, K., Takada, H., Ahmed, S., Obinata, K., Hamada, K., Okimoto, Y., and Hara, T. 2002. Genetic basis of patients with bacille Calmette-Guerin osteomyelitis in Japan: Identification of dominant partial interferon-gamma receptor 1 deficiency as a predominant type. J. Infect. Dis. **185**:706–709.

142. Schorey, J. S., Carroll, M. C., and Brown, E. J. 1997. A macrophage invasion mechanism of pathogenic mycobacteria. Science **277**:1091–1093.

143. Scott, P., Pearce, E., Cheever, A. W., Coffman, R. L., and Sher, A. 1989. Role of cytokines and CD4+ T-cell subsets in the regulation of parasite immunity and disease. Immunol. Rev. **112**:161–182.

144. Sinclair, J., and Sissons, P. 1996. Latent and persistent infections of monocytes and macrophages. Intervirology **39**:293–301.

145. Spector, W. G. 1975. The dynamics of granulomas and the significance of epithelioid cells. Pathol. Biol. (Paris) **23**:437–439.

146. Stevens, J. G., Haarr, L., Porter, D. D., Cook, M. L., and Wagner, E. K. 1988. Prominence of the herpes simplex virus latency-associated transcript in trigeminal ganglia from seropositive humans. J. Infect. Dis. **158**:117–123.

147. Straus, S. E., Ostrove, J. M., Inchauspe, G., Felser, J. M., Freifeld, A., Croen, K. D., and Sawyer, M. H. 1988. NIH conference. Varicella-zoster virus infections. Biology, natural history, treatment, and prevention. Ann. Intern. Med. **108**:221–237.

148. Stupp, Y., Paul, W. E., and Benacerraf, B. 1971. Structural control of immunogenicity. 3. Preparation for and elicitation of anamnestic antibody responses by oligo- and poly-lysines and their DNP derivatives. Immunology 21:595–603.

149. Sturgill-Koszycki, S., Schaible, U. E., and Russell, D. G. 1996. Mycobacterium-containing phagosomes are accessible to early endosomes and reflect a transitional state in normal phagosome biogenesis. EMBO J. 15: 6960–6968.

150. Sweet, C. 1999. The pathogenicity of cytomegalovirus. FEMS Microbiol. Rev. 23:457–482.

151. Tenser, R. B., Hay, K. A., and Edris, W. A. 1989. Latency-associated transcript but not reactivatable virus is present in sensory ganglion neurons after inoculation of thymidine kinase-negative mutants of herpes simplex virus type 1. J. Virol. 63:2861–2865.

152. Thomas, S. K., Gough, G., Latchman, D. S., and Coffin, R. S. 1999. Herpes simplex virus latency-associated transcript encodes a protein which greatly enhances virus growth, can compensate for deficiencies in immediate-early gene expression, and is likely to function during reactivation from virus latency. J. Virol. 73:6618–6625.

153. Thomas, S. K., Lilley, C. E., Latchman, D. S., and Coffin, R. S. 2002. A protein encoded by the herpes simplex virus (HSV) type 1 2-kilobase latency-associated transcript is phosphorylated, localized to the nucleus, and overcomes the repression of expression from exogenous promoters when inserted into the quiescent HSV genome. J. Virol. 76:4056–4067.

154. Tomazin, R., Boname, J., Hegde, N. R., Lewinsohn, D. M., Altschuler, Y., Jones, T. R., Cresswell, P., Nelson, J. A., Riddell, S. R., and Johnson, D. C. 1999. Cytomegalovirus US2 destroys two components of the MHC class II pathway, preventing recognition by CD4+ T cells. Nat. Med. 5:1039–1043.

155. Tort, J., Brindley, P. J., Knox, D., Wolfe, K. H., and Dalton, J. P. 1999. Proteinases and associated genes of parasitic helminths. Adv. Parasitol. 43:161–266.

156. Tsavachidou, D., Podrzucki, W., Seykora, J., and Berger, S. L. 2001. Gene array analysis reveals changes in peripheral nervous system gene expression following stimuli that result in reactivation of latent herpes simplex virus type 1: Induction of transcription factor Bcl-3. J. Virol. 75:9909–9917.

157. Tura, B. J., Bunyan, K. E., and Harrison, D. J. 2001. The effect of IFNgamma on the hepatocyte: Cell cycle and apoptosis. Int. J. Exp. Pathol. 82:317–326.

158. van Rie, A., Warren, R., Richardson, M., Victor, T. C., Gie, R. P., Enarson, D. A., Beyers, N., and van Helden, P. D. 1999. Exogenous reinfection as a

cause of recurrent tuberculosis after curative treatment. N. Engl. J. Med. 341:1174–1179.

159. Villella, A., Picard, C., Jouanguy, E., Dupuis, S., Popko, S., Abughali, N., Meyerson, H., Casanova, J. L., and Hostoffer, R. W. 2001. Recurrent *Mycobacterium avium* osteomyelitis associated with a novel dominant interferon gamma receptor mutation. Pediatrics **107**:E47.
160. Weinrauch, Y., Drugan, D., Shapiro, S. D., Weiss, J., and Zychlinsky, A. 2002. Neutrophil elastase targets virulence factors of enterobacteria. Nature **417**:91–94.
161. Wells, C. L., van de Westerlo, E. M., Jechorek, R. P., and Erlandsen, S. L. 1996. Intracellular survival of enteric bacteria in cultured human enterocytes. Shock **6**:27–34.
162. Wesch, D., Kabelitz, D., Friese, K., and Pechhold, K. 1996. Mycobacteria-reactive gamma delta T cells in HIV-infected individuals: Lack of V gamma 9 cell responsiveness is due to deficiency of antigen-specific CD4 T helper type 1 cells. Eur. J. Immunol. **26**:557–562.

CHAPTER 4

Quantitative and qualitative changes in bacterial activity controlled by interbacterial signalling

Simon Swift

DEFINITIONS OF QUORUM SENSING

Bacteria modulate their phenotype in response to physiochemical changes in their environment. The changes may result from the movement of the bacteria from one location to another, the actions of the bacteria modifying their environment or the responses of a host to the presence of the bacteria. The responses of the bacterium to these changes in its environment are often at the level of the regulation of gene expression. In the study of bacterial pathogenesis it is necessary to ascertain which environmental parameters can signal the virulent phenotype and how the bacterium perceives and responds to these environmental parameters. Some factors are based upon the properties of the host, e.g., temperature (19, 37), iron restriction (86) and the prevailing ionic environment (3, 31, 38, 115). One important parameter is the size of the bacterial population, where changes are regulated by bacterium-to-bacterium signalling in a process that perceives the population cell density, known as quorum sensing (see reviews in 27, 29, 63, 98, 108, 110). It is the combined response to all the sensory input that the bacterium receives that determines the final phenotype, but for many examples quorum sensing provides a dominant signal (see 113 for a review).

In this review the basic concepts of bacterium-to-bacterium signalling are introduced and the mechanisms by which this signalling contributes to the control of bacterial virulence, with special notice taken of the role of signalling in the switching of bacteria from a benign state to a pathogenic state, are described. Examples range from the production of the ultimate in bacterial survival mechanisms – the endospore – to more subtle changes in bacterial physiology.

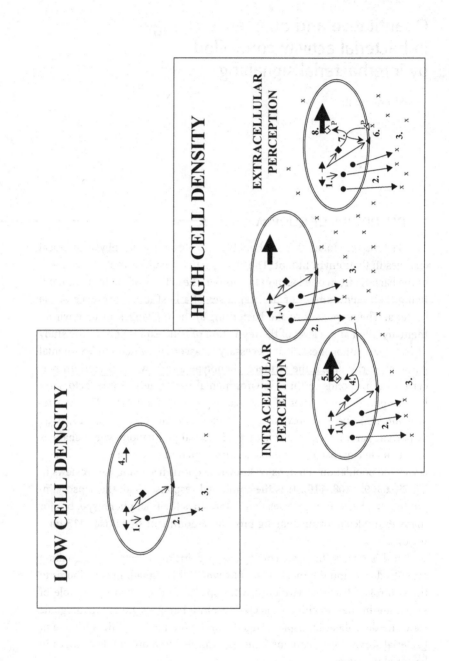

What does dormancy in microbial disease mean in the context of gene regulation by quorum sensing? It is important to look at this question from the perspective of the bacterial population. The properties of a "large" population of bacteria include the ability to produce a significant amount of a toxin or exoenzyme, sufficient to overcome a host, and the ability to rapidly utilise available nutrients. A "small" population may, however, still be present at a high cell density if contained in a small space, e.g., an abscess, a granuloma or as a microcolony growing upon a surface. The coordination of the expression phenotypic characteristics to high cell density populations may be of advantage to the bacterium in deciding upon host-damaging (and immune-response-eliciting) virulence factor expression, the induction of stationary phase survival strategies and the differentiation of cells, e.g., in a mature biofilm. Hence, dormancy for the bacterium may mean persistence in either a highly resistant, (almost) metabolically inactive state, or a pathogenically silent state that evades host defences.

SIGNALS AND SYNTHESIS, PERCEPTION AND RESPONSE

The term quorum sensing was coined in 1994 in a review by Fuqua, Winans, and Greenberg (27), describing the use of N-acylhomoserine lactones by certain Gram-negative bacteria to signal the expression of a number of phenotypic characteristics at high cell density. Since then it has become apparent that other signalling chemistries are used (see 36, 98, 108 for reviews), and the term quorum sensing is now often (if incorrectly) used to describe interbacterial signalling in general. The underlying principle of quorum sensing is simple (see Fig. 4.1); bacteria produce a signal throughout growth, as the population size and density increases, so the concentration of signal increases until a threshold concentration, indicative of a threshold population density, is reached (27, 97). Above this threshold, perception of the signal by the bacterium results in the quorum response, a process that often, but not always (112), involves the modulation of transcriptional activity.

Figure 4.1. Quorum sensing. At low cell density, genes for the biosynthesis, perception and response to the signal are expressed at a basal level (1.). The signal (x) is made and exits the cell (2.), but does not accumulate (3.), and the quorum response is not activated (4.). At high cell density, genes for the biosynthesis, perception and response to the signal are expressed (1.). The signal is made and exits the cell (2.), where it accumulates (3.) and is perceived (4., 6.). If perception is intracellular there is direct activation of the response regulator (4.), e.g., LuxR-type, and the quorum response is activated (5.). If perception is extracellular there is activation of a sensor kinase (6.), phosphotransfer to activate the response regulator (7.), and the quorum response is initiated (8.).

At the molecular level this involves the signal and the cellular machinery involved in signal production, signal perception and the response to the signal. Taking the signal first: it is possible to make a gross subdivision of bacterial signals along the lines of the results of Gram's stain. Signalling in Gram-negative bacteria is often mediated by N-acylhomoserine lactone signals (acyl-HSLs), whereas in Gram-positive bacteria the signals are often peptide-based. Nevertheless, the reader should be aware that the discovery of the widespread nature of bacteria-to-bacteria signalling has generated a significant amount of research interest that has highlighted the presence of many other signal chemistries; fatty acyl methyl esters (23), quinolones (83), cyclic dipeptides (35), furanones (91, 92), indole (105) and others as yet to be determined (for examples see 5, 112). The reader is also directed to the reviews in (45, 46, 63, 98, 108).

Signal generation for acyl-HSLs appears simply to be the coupling of amino acid and fatty acid biosynthesis. More interestingly, it is now apparent that at least three different synthase families exist: the LuxI family (see 30), the LuxM family (see 64) and the HdtS family (49). The primary molecular substrates for this reaction have been determined to be S-adenosyl methionine and acylated acyl carrier protein in a number of independent studies for members of the LuxI family (reviewed in 26) and the AinS protein (32). For peptide signals the ribosomal synthesis of a precursor propeptide is followed by processing, which often introduces other chemical groups (e.g., lipid moieties, as seen with the ComX pheromone of *Bacillus subtilis*, reviewed by (52)), intramolecular bonds (e.g., thiolactone, in the staphylococcal autoinducing peptide AIP (40, 60)) and cleaves the precursor to release the mature peptide. Is signal generation a regulatory step? In many cases the expression of signal generators forms part of the quorum response, providing positive feedback that allows a rapid induction of the high cell density phenotype. For some signals substrate availability may coordinate signal production with nutrition, although there is little evidence to suggest that this is a widespread strategy.

How does the signal exit the cell? In the case of acyl-HSL molecules with short acyl chains the freely diffusible nature of these molecules has been demonstrated (43, 74). Acyl-HSLs with longer acyl chains do not appear to escape the cell membranes as easily and for N-3-oxododecanoylhomoserine lactone (3-oxo-C12–HSL) produced in *Pseudomonas aeruginosa* at least, the signal is actively pumped from the bacterial cell (74). For peptide signals active export is the norm, with ATP-binding cassette (ABC) transporters frequently used (e.g., for AIP (*Staphylococcus aureus*), CSP (competence stimulating peptide,

Streptococcus pneumoniae), CSF (competence and sporulation factor, also termed the Phr pheromones, which is Sec dependent; (94), *B. subtilis*), Nisin (*Lactococcus lactis*) and possibly ComX (*B. subtilis*); see 46, 52, 62 for reviews).

Perception of the signal can be accomplished by receptors externally (surface exposed receptors) or internally situated (see Fig. 4.1). Taking the two major classes of signal, acyl-HSLs and peptides, examples of both internal and external sensing are apparent. Perception of acyl-HSLs by LuxR family response regulators is internal. In the case of *Agrobacterium tumefaciens* the LuxR ortholog TraR perceives the acyl-HSL signal as a monomer on the inner face of the inner cytoplasmic membrane (85). Holo-TraR dimerises and is cytoplasmic, where it acts as a transcriptional activator for the quorum response (85). It is however unclear, as yet, whether all LuxR family proteins act in this way. Response to acyl-HSLs produced by *Vibrio harveyi* is through de-repression of the *luxCDABEGH* operon via a phosphotransfer relay, allowing transcriptional activation to occur (as reviewed by 63, 108). A model has been proposed where, at low cell densities, the activation of an unknown repressor of the *lux* operon is mediated by the phosphorylated form of the regulatory protein LuxO and an alternative sigma factor, σ^{54} (53). Phosphorylation of LuxO is in part dependent upon the presence or absence of acyl-HSL molecules and the effect they have on their cognate response regulator, LuxN (21, 22). The majority of peptide signals are perceived by sensor kinase proteins, which generally activate transcriptional activators of the quorum response (see 46 for a review). The exception is the Phr pheromones of *B. subtilis* that enter the cell through the oligopeptide permease (OPP; 80, 88), are perceived internally by Rap phosphatases (which they inhibit) and thereby influence the quorum response by affecting the level of phosphorylated transcriptional activators (see 78 for a review).

PEPTIDE SIGNALLING IN *BACILLUS SUBTILIS* AND THE CONTROL OF ENDOSPORE FORMATION

Peptide signalling in *Bacillus subtilis* mediates the population density control of sporulation (see 79, 96 for reviews), the development of genetic competence and other physiological processes including the production of the biosurfactant surfactin and the lantibiotic subtilin (see Fig. 4.2; see 47, 52 for reviews). Growth phase regulation of subtilin is autoregulatory (47), whereas the regulation of sporulation and genetic competence and surfactin production are linked in a process involving sensory inputs perceiving cell density and nutrient availability (52, 78, 96).

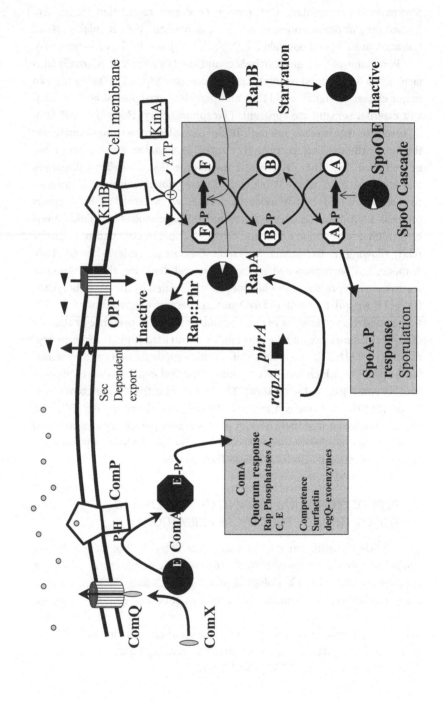

The bacterial endospores are the most resistant living structures known (68). The decision committing a bacterium to the energy-expensive sporulation process involves a complex regulatory network that allows the sequential coordination of the necessary events. Central to the initiation of this process are secreted peptide signals.

The ComX peptide is a general indicator of high cell density in *Bacillus* populations (58). The *comQXPA* genes form a signalling cassette in *Bacillus*, which is homologous to those found in a number of other Gram-positive bacteria, including the Agr signalling cassette of *Staphylococcus aureus*, the Com signalling cassette of *Streptococcus pneumoniae* and the Pln signalling cassette of *Lactobacillus plantarum* (52). The *comX* open reading frame (ORF) encodes a 55–amino acid peptide, which is processed for export leaving the C-terminal 10 amino acids with an, as yet, uncharacterised lipidation on a tryptophan residue (58). ComQ is thought to participate in this process, and no mutants have been identified that affect ComX pheromone production other than those in *comQ* or *comX* (52). Exactly how the peptide exits the cell is unknown, although analysis of the *B. subtilis* genome has identified a number of peptide ABC transporter proteins with homology to export systems for peptides in other organisms (48, 52). The histidine sensor kinase encoded by *comP* mediates the perception of ComX pheromone, and phosphotransfer from activated phospho-ComP to ComA results in the quorum response (competence, surfactin, regulation of degradative enzymes, and Rap phosphatases) activated by phopsho-ComA (reviewed in 51, 52).

What does this have to do with dormancy and sporulation? Sporulation is controlled by the opposing activities of a distinct family of histidine kinases (KinA–E) and the Rap family of phosphatases (41). Peptide (ComX and Phr) signalling has a central role in this control. The key regulator in the initiation of sporulation is SpoOA (10). High levels of phosphorylated SpoOA are required to decide for sporulation (4, 8). Phospho-SpoOA levels are determined

Figure 4.2. Peptide signalling in *Bacillus*. The ComA transcriptional activator mediates the expression of the ComA quorum response (see box) in the presence of the modified ComX pheromone. ComA is activated by phosphorylation by the signal sensor kinase, ComP. Sporulation occurs in the presence of sufficient phospho-SpoOA. The level of phospho-SpoOA is a balance between input phosphorylation signals (e.g., via KinA, KinB) and phosphatase activity (e.g, Rap, SpoOE). RapB phosphatase activity is high in actively growing cells but is inactivated by starvation. RapA, C and E phosphatases are induced as part of the ComA quorum response and inactivated by the Phr peptide signals after export, processing and import.

by the balance between the activity of an activating kinase cascade and the inactivating phosphatases. SpoOF acquires phosphorylation from any of KinA through to KinE and transfers this to SpoOA via SpoOB. The level of phospho-SpoOA is kept low by the activity of specific phosphatases (of the Rap family and SpoOE) acting on the phosphorylated members of the activating cascade and also on phospho-SpoOA itself (42, 82). The regulation of the transcription of *spoOA* and other members of the SpoOA phosphotransfer cascade is also an important factor (84, 107).

An analysis of the *B. subtilis* genome sequence has identified 11 Rap family phosphatases. Seven of the *rap* genes are followed by small ORFs encoding a precursor for putative signalling peptides (79). The peptides are processed and exported via the SecA-dependent system (41) and re-imported via the oligopeptide permease (Opp; 80, 88) system. Interaction of the pentapeptide Phr signal with its target Rap phosphatase inhibits activity and thereby allows accumulation of phospho-SpoOA and thus sporulation (41, 77). With the exception of *phrA*, each peptide ORF is preceded by a sigmaH-(starvation) dependent promoter (79). At least three of the *rap* genes are activated by phospho-ComA, making RapA, RapC and RapE part of the ComA quorum response. RapB transcription is activated by conditions favouring vegetative growth (41, 65, 81). Hence, sporulation will not be initiated in conditions favouring either (1) vegetative growth, as a high level of RapB activity (dephosphorylating phospho-SpoOF) ensures low levels of phospho-SpoOA or (2) the development of genetic competence as high-level RapA (dephosphorylating phospho-SpoOF) and also RapC and RapE activities will ensure low levels of phospho-SpoOA. It is only when conditions of starvation occur that we see an increased level of Phr peptide production and inhibition of the dephosphorylation activities coupled with increased input kinase activity from the KinA–E family that phospho-SpoOA accumulates and sporulation begins.

We therefore see a dynamic interaction between multiple peptide and phosphotransfer signalling systems that provides numerous checkpoints before commitment to dormancy. Is the peptide signalling involved here true quorum sensing control? The exported peptides are only detected in the culture medium of import defective *opp* mutants, and given the extent of exoprotease production by *Bacillus* it is thought unlikely that exported peptides will exhibit a long extracellular half-life. The hypothesis is therefore that signalling through the control and timing of export, processing and re-import provides a self-signalling or an immediate microenvironment signalling activity rather than a population-wide signalling event (77).

What does this have to do with virulence and dormancy? *B. subtilis* is not noted for its pathogenicity, however there are numerous members of the genera of *Bacillus* and *Clostridium* that can lie dormant as spores for many years before germination at host sites cause disease. Armed with this wealth of information for *B. subtilis* it is now important to determine just how much of the signalling control is used by pathogenic spore-forming bacteria, especially as the genome sequence of the non-pathogenic *Clostridium acetobutylicum* has no orthologs of either KinA to KinE or the Rap phosphatases (69).

BACTERIAL CYTOKINES IN *MICROCOCCUS* AND *MYCOBACTERIUM* CONTROL THE EXIT FROM DORMANCY

A third of the world's population is infected with *Mycobacterium tuberculosis*, the causative agent of tuberculosis. In 2000, approximately 10% of these (8.4 million) developed the disease and 1.5 million died. It is estimated that a further 0.5 million died as a consequence of HIV/AIDS and tuberculosis together (24, 114). The treatment of *Mycobacterium tuberculosis* infections is complicated by the rise of multidrug resistant strains coupled with the problems related to the slow growth rates and dormancy of the organism, which impede diagnosis and mean that drug treatments are required over long time periods for efficacy. The World Health Organisation's figures quoted above can be regarded in a different way; in 90% of infections (75 million people!) the infection is latent and the infective bacteria are dormant. Understanding the biology of dormancy, therefore, appears as an attractive target and may lead to better diagnostic procedures and the possibility of new drugs that can control bacteria in the dormant state.

Following the ingestion of *M. tuberculosis* by macrophages the majority of bacteria are initially found in fused phagolysosomes. Virulent strains are able to resist killing by reactive Nitrogen species and proliferate, although they may suffer a transient injury. *M. tuberculosis* persists in macrophages within a granuloma in the organs of infected hosts (see 24, 87 for reviews). The granuloma consists of macrophages, giant T cells, T cells, B cells and fibroblasts. In latent infections the bacteria are contained in a dormant state by the immune system, but not cleared. Breakdown of the immune responses containing the infection reactivate the bacteria, which begin to replicate, with associated tissue damage. (See Chapter 7 for a full discussion of dormant *M. tuberculosis*.)

The discovery of interbacterial communication between cells of a high G+C bacterium, *Micrococcus luteus*, phylogenetically related to *M. tuberculosis* may provide a crucial step forward in improving the diagnosis and treatment of tuberculosis (see 45 for a review). *M. luteus* cells enter a recognised state of persistent dormancy upon the entry into stationary phase (44). Exponentially growing cells have a viability/cultivability approaching 100%, i.e., micro-scopic cell counts equate to colony-forming units (CFU) that will grow on agar plates or the viable count from the most probable number (MPN) method. Replication of the experiment with dormant cells gives a reduction in cul-tivability of 10,000 fold. However, in the presence of sterile filtered culture supernatant from late logarithmic cultures viability in MPN counts can be restored to approaching 100% (66). Analysis of the culture supernatant re-vealed that the active molecule was a protein of 16 to 17 kDa termed Rpf (resuscitation-promoting factor; 66). Protein microsequencing of N-terminal and internal peptides from Rpf allowed a reverse genetic approach to cloning the *rpf* gene. Database searches found homologues to the predicted Rpf se-quence in other high G+C bacteria (*Mycobacterium, Corynebacterium, Strep-tomyces*). Five homologues were found in the genomic sequence of *M. tuber-culosis*. Moreover, in the presence of recombinant micrococcal Rpf (rmRpf; produced in *Escherichia coli*) the improved cultivability of a number of species of both rapid- and slow-growing mycobacteria was demonstrated (7, 66).

To further examine the importance of this signalling in disease the effect of rmRpf upon *M. tuberculosis* from murine infections was investigated. Mi-croscopic and surface property analyses of bacteria isolated from macrophage lysates from infected mice have shown a progressive change in the morphol-ogy of the bacteria from short rods, (presumably) through ovoid forms, to coccoid forms (7). Significant alterations in the cell envelope are observed following prolonged exposure to the intracellular macrophage environment, including the reduction in bacteriophage adsorption and bacterial clumping and the acquisition of ethidium bromide staining after heat treatment. There is also a significant and highly variable reduction in cultivability. Microscopic analysis of macrophage suspensions taken from infected mice reproducibly demonstrated approximately 10^6 mycobacteria per ml. Nevertheless, CFU (determined by plating on agar plates) or MPN (determined by dilution into liquid media) counts estimated the number of viable cells in these samples to be between zero and 10^4 per ml (7). In these experiments the CFU counts were particularly low. Are the bacteria that were not cultivable here dormant? (See Chapter 1 for further information about cultivability in general and Chapter 7 for information on *M. tuberculosis* in particular.)

In vitro tests show that for actively growing mycobacteria the CFU count is comparable to the MPN count; however, as cultures are left to age, a decreas-ing number of bacteria are able to form colonies. Cultivability is enhanced in the MPN method and is almost equivalent to the total count when MPN is determined in the presence of picomolar concentrations of rmRpf (7). Extend-ing this study to the culturing of both virulent and avirulent bacteria from macrophage lysates from infected mice, the presence of rmRpf in MPN assays gave a 10- to 20-fold increase in cultured bacteria from the macrophages (7). The conclusion drawn is that there are cultivable bacteria, injured bacteria resuscitable with rmRpf and injured or dead bacteria that are unable to be resuscitated with rmRpf.

What is the significance of this work? Although the effect of rmRpf on infection-derived bacteria is not as impressive as that upon laboratory grown cultures it is clear that there is an effect. The challenge now is to determine the biological role of the five mycobacterial Rpf-like proteins in infections. It is predicted that four will be secreted and one will be anchored in the cell envelope (7, 45). It is suggested that phagosomal enzymes will inactivate these Rpf-like proteins, thus providing a simple explanation for both the poor cultivability of macrophage-derived mycobacteria and their responsiveness to rmRpf. In this case the activity of mycobacterial Rpf-like proteins is necessary for bacterial proliferation and we may expect to see further improvements upon the cultivability if mycobacterial Rpf-like proteins, rather than rmRpf, are used. This may have a positive impact upon both the diagnosis of mycobacterial disease and perhaps form a target for useful anti-mycobacterials. If this is the case, a knowledge of the conditions that induce expression of Rpf-like proteins in the host will be important as it may be that Rpf-like proteins influence the ability to escape from immune containment and to cause disease. Moreover, why are there five such proteins? It may be that they are differentially expressed giving different responses for a range of conditions. Or it may simply be a case of functional redundancy, with the presence of five gene copies indicating the importance of their function? Whatever is the case there are many questions for the future; for example, how do mycobacteria perceive the cytokine, what are the responses to the cytokine, and is there an interaction with, or cross talk with the (human) host.

THE INHIBITION OF REPLICATION IN *ESCHERICHIA COLI*

In dormancy bacteria shut down metabolic activity to a minimum and will cease DNA replication and cell division. One mechanism by which bacterium-to-bacterium signalling may act, therefore, is to contribute to the control of these processes. A role for quorum sensing in the control of cell division via the activation of the *ftsQ* promoter by the LuxR homologue SdiA has been suggested, tested, but never proven for *Escherichia coli* (28, 95, 104). In *Salmonella* it has been shown that two classes of promoters respond to SdiA. The first is induced in the presence of chromosomal *sdiA* and acylhomoserine lactones (possibly produced by competitor species), the second by SdiA over-expression (plasmid-borne *sdiA*) alone. The *ftsQ* promoter falls in the second, potentially artefactual class (61). The evidence for an involvement in replication came from a study of the initiation of replication in *Escherichia coli* (112). The thermosensitive replication mutant MG1655*dnaC2* will not initiate replication at a non-permissive temperature (111). The ability of this strain to re-initiate replication after shifting back to the permissive temperature is dependent upon population cell density. At optical density above 0.4, re-initiation does not occur (111). An inhibitory component of cell free media conditioned by bacterial growth was shown to be responsible for the effect, although the structure of this compound remains unknown (112). Experiments demonstrated that the block was most likely at the initiation of replication and not at elongation as the addition of concentrated conditioned medium (cCM) to bacterial cells after initiation did not prevent the completion of chromosomal replication (112). The molecular biology of the signalling events described by Withers and Nordström (112) is significantly different from that described elsewhere in that the quorum response is independent of transcription and translation. *Escherichia coli* cells are able to initiate replication in the absence of transcription and translation; however, the presence of the factor in cCM can block this. The molecular events defining signalling, perception and response with respect to the initiation of replication are therefore intriguing and may provide insight into the control of dormancy in a number of bacterial species.

PATHOGENICALLY DORMANT BACTERIA

In some cases bacteria can be present and metabolically active but not actively virulent until circumstances favour pathogenesis. Put simply, some bacteria will lie low and try to proliferate while evading host defences until such a time as the population has gathered sufficient numbers to produce a pathogenic insult capable of overwhelming the host and its defences. These bacteria can be considered pathogenically dormant but not metabolically dormant. The roles of population size, evasion of host defence and quorum sensing are entwined in the control of pathogenicity of at least two important pathogens, *Staphylococcus aureus* and *Pseudomonas aeruginosa*. Both organisms are common in our environment and are responsible for a wide range of infections (see 2, 15, 55, 56, 101 for reviews). Often these infections are hospital-acquired and are difficult to treat because of antibiotic resistance (reviewed by 15, 55, 106). The pathogenesis of both species relies upon the coordinated expression of multiple virulence factors, a process in which quorum sensing, via acyl-HSLs for *P. aeruginosa* and via modified peptides for

S. aureus, has a central role (39, 70, 90, 101, 110). Allied with this is the capacity of both organisms to form infection-related biofilms (15, 16, 34). A biofilm is a persistent mode of growth at a surface within a polymeric matrix exhibiting a resistant physiology. The bacterial cells within a biofilm are at high cell densities, and cell-to-cell signalling has been shown to play a central regulatory role in the development of a mature, resistant biofilm (17, 18, 33, 103). (See Chapter 6 for more information about biofilms.)

Quorum Sensing is Essential for the Full Virulence of *Pseudomonas aeruginosa*

Pseudomonas aeruginosa employs a multilayered hierarchical quorum-sensing cascade that links Las-signalling (LasR/LasI/3-oxo-C12-HSL), Rhl-signalling (RhlR/RhlI/*N*-butanoylhomoserine lactone [C4-HSL]), quinolone-signalling (PQS) and the alternative sigma factor RpoS to integrate the regulation of virulence determinants and the development of persistent biofilms with survival under environmental stress (see 56, 73, 90, 98). The quorum response of *P. aeruginosa* is summarized in Fig. 4.3, where it can be seen to coordinate the activation of the major virulence determinants. The quorum response can be subdivided into genes (i) that are induced only by 3-oxo-C12-HSL, (ii) that are induced only by C4-HSL, (iii) that are induced either by C4-HSL or 3-oxo-C12-HSL, and (iv) that are only induced by C4-HSL and 3-oxo-C12-HSL (109). The Las- and Rhl-signalling systems have also been implicated in the regulation of exoenzyme secretion via the general secretory pathway (11). Control of secretion by quorum sensing presumably enables *P. aeruginosa* to deal with the enormous induction of exoproduct (e.g., elastase, LasA protease, exotoxin A) synthesis as the bacterial population reaches a high cell density.

More importantly, it has been possible to show that acyl-HSL signalling is essential for the development of full virulence by *P. aeruginosa* during an infection. The effect of specific mutations in *rhlI*, *lasR* and *lasI* has been investigated in murine models of acute pulmonary infections and burn wound infections (75, 89, 99). In the burn wound model *lasR*, *lasI* and *rhlI* mutants are significantly less virulent than the parent *P. aeruginosa* strain PAO1 (89). A third-degree (full thickness) burn was produced by exposure of the shaved skin to 90° C water for 10 seconds and subjected to a subcutaneous challenge of 2×10^2 to 3×10^2 CFU *P. aeruginosa*. After 48 hours the wild-type strain shows an average mortality of 94%. In contrast, the mortality accredited to signalling mutants is significantly reduced: *lasR*, 28%; *lasI*, 47%; *rhlI*, 47%; and *lasI*, *rhlI* double mutant, 7%. The virulence of the mutants was restored

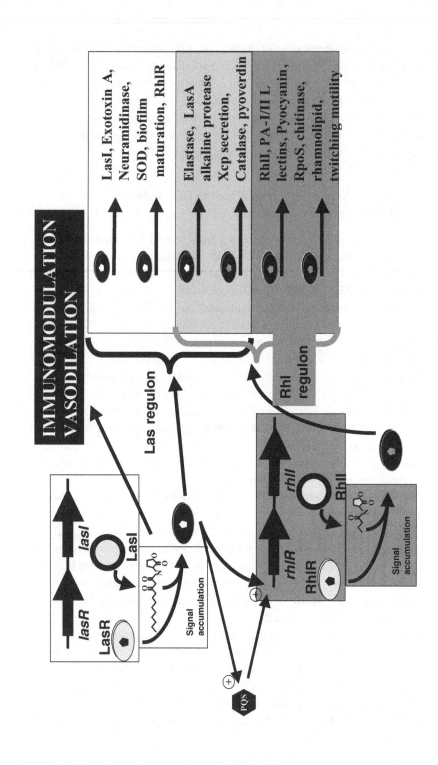

IMMUNOMODULATION VASODILATION

LasI, Exotoxin A, Neuramidinase, SOD, biofilm maturation, RhlR

Elastase, LasA alkaline protease Xcp secretion, Catalase, pyoverdin

RhlI, PA-I/II L lectins, Pyocyanin, RpoS, chitinase, rhamnolipid, twitching motility

Las regulon

Rhl regulon

lasR lasI
LasR LasI

Signal accumulation

rhlR rhlI
RhlR RhlI

Signal accumulation

PQS

by complementation with plasmids expressing LasI, RhlI or LasI and RhlI, although the *lasR* mutation could not be complemented, as the plasmid was not stably maintained. To assess the spread of *P. aeruginosa* within the burned skin, bacterial counts were made at the site of inoculation and at a site 15 mm distant. At the inoculation site cell numbers showed no significant variation between parent and mutant strains. At the distant site, however, although the single *rhlI* and *lasI* mutations had no significant effect upon the spread of the bacteria, mutants with defects in *lasR* or both *rhlI* and *lasI* showed no spread to the distant site until after 16 hours from inoculation. The analysis of *lasR* and *lasI rhlI* double mutant bacteria recovered from the distal site demonstrated that the effect was not due to suppression of any of the mutations as observed by Van Delden *et al.* (102). These data suggest that although there is some redundancy in the control of the important virulence factors via *las* and *rhl* signalling, quorum sensing is necessary for the optimal coordination of virulence factor expression for pathogenicity.

A similar situation is apparent in the pulmonary infection model. Neonatal mice were intranasally challenged with 1.5×10^9 CFU of *P. aeruginosa* and sacrificed after 24 hours. For each mouse, pneumonia was defined by the presence of $>10^3$ CFU of *P. aeruginosa* in a lung homogenate and by histopathological evidence of destruction of the lung parenchyma, oedema and leukocyte infiltration. Of the mice inoculated with the parental strain, 55% developed confluent pneumonia throughout the lungs, with a mortality rate of 21% of the inoculated animals. In contrast only 10% of mice inoculated with a *rhlI lasI* double mutant developed pneumonia, and this was much less severe than that seen with the parent strain. Full virulence could be restored to the double mutant by complementation of the *rhlI lasI* mutations with plasmid borne copies of *rhlI* and *lasI* (75). In agreement with a role for signalling in pulmonary infection Tang *et al.* (99) demonstrated that although a *lasR* mutant could colonise the murine lung, it was avirulent,

Figure 4.3. *N*-Acylhomoserine lactone signalling in *Pseudomonas aeruginosa*. A simplified model of the hierarchical Las/Rhl quorum-sensing system of *P. aeruginosa* is shown. LasR activates transcription of LasI and the signal 3-oxo-C12-HSL is generated. Interception of this signal by the host has immunomodulatory and vasodilatory effects. LasR/3-oxo-C12-HSL activates the Las quorum response, which includes the virulence-associated genes in the Las-regulon box, the genes for the synthesis of the PQS signal and RhlR. RhlR activates transcription of *rhlI* (also positively influenced by PQS); RhlI produces C4-HSL and expression of the overlapping Rhl quorum response, which includes the virulence and survival-associated genes in the Rhl-regulon box, is activated by RhlR/C4-HSL.

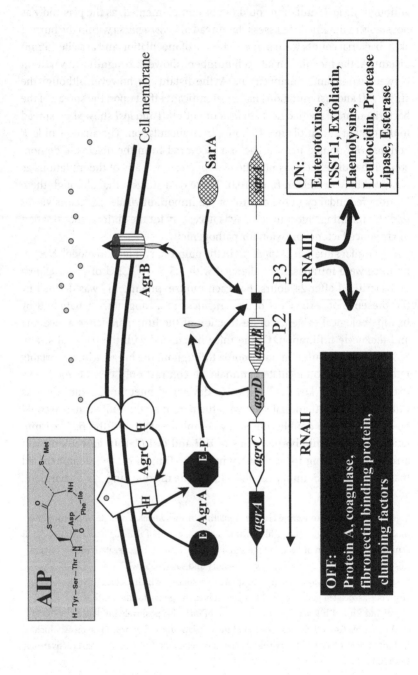

AIP

H–Tyr–Ser–Thr–Cys–Asp–Phe–Ile

Cell membrane

AgrB

SarA

AgrC

AgrA

E-P

E

P3

RNAIII

P2

RNAII

agrA   *agrC*   *agrD*   *agrB*   *sarA*

ON:
Enterotoxins,
TSST-1, Exfoliatin,
Haemolysins,
Leukocidin, Protease
Lipase, Esterase

OFF:
Protein A, coagulase,
fibronectin binding protein,
clumping factors

being unable to achieve high cell densities, cause pneumonia or penetrate into deeper tissues.

Las-signalling is not restricted to within the bacterial population as the cells of the mammalian host can intercept the signals. The immunosuppressive effect of 3-oxo-C12-HSL upon the host immune system observed by Telford *et al.* (100) prejudices against antibacterial action, and so favours the bacteria. Additionally the vasorelaxant activity of 3-oxo-C12-HSL observed by Lawrence *et al.*, (50) may contribute to the ability of the bacteria to maintain the supply of key nutrients to the site of infection by increasing local blood flow. Hence, in addition to coordinating the expression of *P. aeruginosa* virulence determinants, Las-signalling may also induce eukaryotic cells to maximize the provision of nutrients via the blood without activation of the host's defence system.

Quorum Sensing is Essential for the Full Virulence of *Staphylococcus aureus*

The coordination of virulence factor expression by *Staphylococcus aureus* involves two major global regulatory systems, Agr and Sar (13, 76). During the initial, low population density stages of a staphylococcal infection, the expression of surface proteins binding extracellular-matrix molecules, e.g., fibronectin, collagen and fibrinogen, and to the Fc region of immunoglobulin, i.e. Protein A, is favoured. This promotes evasion of host defences and the successful colonisation of host tissues. (see 25, 67 for reviews). *Staphylococcus aureus* challenges the host immune system by eliciting a regional inflammation and subsequent abscess formation. Inside the effectively closed system of the abscess, bacterial population density increases and secreted enzymes and toxins are induced that efficiently destroy white blood cells and liberate nutrients from tissue. Quorum sensing in *S. aureus* coordinates the downregulation of surface-associated factors and the induction of exoenzymes and toxins (see 70 for a review).

The Agr locus, consisting of two divergent operons, P2 and P3, is central to quorum-sensing control in *S. aureus* (Fig. 4.4). The P2 operon comprises

Figure 4.4. Peptide thiolactone signalling in *Staphylococcus aureus*. The *agr* signalling cassette encodes AgrB, which participates in the export and processing of the autoinducing peptide (AIP) from the precursor encoded by *agrD*. AgrC perceives the threshold concentration of AIP, which induces autophosphorylation and phosphotransfer to the transcriptional activator AgrA. Phospho-AgrA activates transcription of the P2 (*agrBDCA*) and P3 (RNAIII) operons in concert with SarA. RNAIII mediates the quorum response.

the *agrBDCA* signalling cassette that is homologous to *comQXPA* of *B. subtilis* (52, 70, 72). P3 encodes the RNAIII molecule that acts as an intracellular signal controlling the transcription of genes within the Agr regulon (6, 71). AgrD encodes a small peptide that is cleaved and processed in a process that involves AgrB and which results in the secretion of a cyclic thiolactone octapeptide or autoinducing peptide (AIP; 40). AIP perception is by the sensor kinase AgrC (39, 54, 57). Molecular events proceed following perception to induce AgrC autophosphorylation on a conserved histidine residue and the subsequent phosphotransfer to an aspartate residue on AgrA (54, 57). Phospho-AgrA acts in conjunction with SarA (12, 14) to activate transcription of P2 (a positive feedback, autoinducing loop) and P3 operons. The intracellular signal RNAIII then activates transcription of the exoenzyme and toxin genes while downregulating transcription of the surface associated factors.

Analogous to the effects of mutations in Las- and Rhl-signalling described above for *P. aeruginosa*, mutations which compromise Agr-signalling reduce the virulence of staphylococci in animal models for arthritis, endophthalmitis, osteomyelitis and skin abscess formation (1, 9, 29, 59). In a Swiss mouse arthritis model, a single tail injection of 10^7 bacteria, *agr*+ or *agrA*::Tn*551*, was administered at 4–6 weeks of age, and the prevalence and severity of arthritis was monitored up to 21 days when the animals were sacrificed for histopathology (1). The results obtained showed that the *agrA* mutants were severely compromised in their ability to cause arthritis after intravenous injection. Three possible reasons for this were suggested: (1) that in the *agrA* mutant the upregulation of cell-associated proteins involved with binding plasma constituents (immunoglobulins, vitronectin, fibronectin) promotes phagocytosis and/or immobilisation in organs other than joints. This was supported by the finding that while the parental strain was found in the joints of the majority of inoculated mice, but not in the spleen, the opposite was true for the *agrA* mutant, which was often found in the spleen of infected animals, but not in the joints. (2) The reduced production of exoenzymes by the *agrA* mutant reduced the ability of the bacteria to penetrate the joint. (3) The ability to bind bone sialoprotein, which is downregulated in the *agrA* mutant, is a crucial virulence factor for the colonisation of the joint.

A similar trend was observed in a rabbit model for osteomyelitis, where the introduction of an *agrA*::Tn551 mutation into an osteomyelitis clinical isolate reduced the incidence and severity of disease; however in this study the mutation did not eliminate the ability to colonise bone or cause histopathological evidence of disease (osteomyelitis; 29). In this model either 2×10^5

or 2×10^6 CFU were microinjected directly into the center of the medullary canal of the radius in the rabbit forelimb after surgery. The results obtained with parental strain at both doses were significantly different from those obtained with no bacterial inoculation. However, the results obtained with the *agrA* mutant were only significantly different from the no-bacteria controls at the higher infection dose. The infection was significantly more severe with the parental strain, when compared to the *agrA* mutant.

In a rabbit endophthalmitis model intravitreal injection of 10 or 1000 CFU of *S. aureus* or an isogenic *agrA* mutant was followed by observations of neuroretinal function and histopathology (9). The contribution of exoenzymes and toxins to this disease was reflected by the significant increase in onset time and the significant decrease in the severity of the disease seen in the rabbits inoculated with the *agrA* mutant when compared with those inoculated with the parental strain. Histopathologic examination 36 hours post-inoculation of eyes infected with the parental strain demonstrated a focal retinal destruction and mild vitritis, whereas eyes infected with the *agrA* mutant remained normal. Nevertheless for both parental and mutant strains the intravitreal growth rate *in vitro* and *in vivo* was not affected.

More importantly, the analysis of Agr-signalling throughout a number of *S. aureus* and other staphylococci has correlated the secretion of divergent AIP signals (that have been classed into activity groups I to IV) with the observation of virulence antagonism between strains of staphylococci (40, 60, 70, 93) and the subsequent testing of AIP antagonists able to produce quorum-sensing blockade as potential anti-staphylococcal pharmaceuticals (57, 59, 60). The principle of control using quorum-sensing blockade relies upon either rendering or maintaining the infecting bacterium as pathogenically dormant, thus allowing more time for conventional antibiotics and/or host immune responses to control the infection (57, 60). In studying the ability of a group I AIP producing *S. aureus* to produce a skin abscess, Mayville *et al.* (59) observed that following subcutaneous injection of approximately 10^8 CFU into the flank of a 6- to 8-week-old hairless mouse a substantial lesion (mean area approximately 900 mm^2) developed and was still present five days post-inoculation (mean area approximately 600 mm^2). The ability of the *S. aureus* was significantly reduced (lesion sizes <50 mm^2) where an *agr* null mutation was introduced or where the parental strain was co-inoculated with either 5 or 10 μg of an interfering peptide (a synthetic version of a group II strain AIP). Subsequent to these findings structure/activity studies with synthetic peptides have begun to determine the key features of the group I signalling peptide with the ultimate aim of producing a peptide able to block Agr-signalling in all or most signalling classes (57, 60).

CONCLUDING REMARKS

The virulence of an infecting bacterium often depends upon whether the bacterium decides to grow and to actively seek nutrients or whether the bacterium decides to lie low and wait for a more opportune time to begin (rapid) proliferation. One of the important environmental signals bacteria sense is the size of their own population, and it is hypothesised that the balance between the nutrient requirements and the nutrient-acquiring capabilities of the bacterial population decides which course is taken. A knowledge of the bacterium-to-bacterium signalling events that occur during an infection now gives opportunities for both novel diagnostics and novel treatments for persistent infections that involve understanding the dormant or latent phases of bacterial growth and give hope for new strategies for the control of infections of bacteria for which the antibiotic options are fast diminishing.

ACKNOWLEDGEMENTS

I am indebted to Paul Williams and the late Gordon S. A. B. Stewart for their personal support, encouragement and advice throughout my research career. I am also grateful to the numerous other scientists who have been part of the quorum-sensing research group at the University of Nottingham, UK for many useful discussions.

REFERENCES

1. Abdelnour, A., Arvidson, S., Bremell, T., Ryden, C., and Tarkowski, A. 1993. The accessory gene regulator (*agr*) controls *Staphylococcus aureus* virulence in a murine arthritis model. Infect. Immun. **61**:3879–3885.
2. Archer, G. L. 1998. *Staphylococcus aureus*: A well-armed pathogen. Clin. Infect. Dis. **26**:1179–1181.
3. Atlung, T., and Ingmer, H. 1997. H-NS: A modulator of environmentally regulated gene expression. Mol. Microbiol. **24**:7–17.
4. Baldus, J. M., Green, B. D., Youngman, P., and Moran, C. P., Jr. 1994. Phosphorylation of *Bacillus subtilis* transcription factor Spo0A stimulates transcription from the *spoIIG* promoter by enhancing binding to weak 0A boxes. J. Bacteriol. **176**:296–306.
5. Barber, C. E., Tang, J. L., Feng, J. X., Pan, M. Q., Wilson, T. J., Slater, H., Dow, J. M., Williams, P., and Daniels, M. J. 1997. A novel regulatory system required for pathogenicity of *Xanthomonas campestris* is mediated by a small diffusible signal molecule. Mol. Microbiol. **24**:555–566.
6. Benito, Y., Kolb, F. A., Romby, P., Lina, G., Etienne, J., and Vandenesch, F. 2000. Probing the structure of RNAIII, the *Staphylococcus aureus agr*

regulatory RNA, and identification of the RNA domain involved in repression of protein A expression. RNA **6**:668–679.

7. Biketov, S., Mukamolova, G. V., Potapov, V., Gilenkov, E., Vostroknutova, G., Kell, D. B., Young, M., and Kaprelyants, A. S. 2000. Culturability of *Mycobacterium tuberculosis* cells isolated from murine macrophages: A bacterial growth factor promotes recovery. FEMS Immunol. Med. Microbiol. **29**:233–240.

8. Bird, T. H., Grimsley, J. K., Hoch, J. A., and Spiegelman, G. B. 1993. Phosphorylation of SpoOA activates its stimulation of in vitro transcription from the *Bacillus subtilis spoIIG* operon. Mol. Microbiol. **9**:741–749.

9. Booth, M. C., Atkuri, R. V., Nanda, S. K., Iandolo, J. J., and Gilmore, M. S. 1995. Accessory gene regulator controls *Staphylococcus aureus* virulence in endophthalmitis. Invest. Ophthalmol. Vis. Sci. **36**:1828–1836.

10. Burbulys, D., Trach, K. A., and Hoch, J. A. 1991. Initiation of sporulation in *B. subtilis* is controlled by a multicomponent phosphorelay. Cell **64**:545–552.

11. Chapon-Hervé, V., Akrim, M., Latifi, A., Williams, P., Lazdunski, A., and Bally, M. 1997. Regulation of the *xcp* secretion pathway by multiple quorum-sensing modulons in *Pseudomonas aeruginosa*. Mol. Microbiol. **24**:1169–1178.

12. Cheung, A. L., and Projan, S. J. 1994. Cloning and sequencing of *sarA* of *Staphylococcus aureus*, a gene required for the expression of *agr*. J. Bacteriol. **176**:4168–4172.

13. Cheung, A. L., Koomey, J. M., Butler, C. A., Projan, S. J., and Fischetti, V. A. 1992. Regulation of exoprotein expression in *Staphylococcus aureus* by a locus (*sar*) distinct from *agr*. Proc. Natl. Acad. Sci. USA **89**:6462–6466.

14. Chien, Y., and Cheung, A. L. 1998. Molecular interactions between two global regulators, *sar* and *agr*, in *Staphylococcus aureus*. J. Biol. Chem. **273**:2645–2652.

15. Costerton, J. W. 2001. Cystic fibrosis pathogenesis and the role of biofilms in persistent infection. Trends Microbiol. **9**:50–52.

16. Cucarella, C., Solano, C., Valle, J., Amorena, B., Lasa, I., and Penades, J. R. 2001. Bap, a *Staphylococcus aureus* surface protein involved in biofilm formation. J. Bacteriol. **183**:1888–1896.

17. Davies, D. G., Parsek, M. R., Pearson, J. P., Iglewski, B. H., Costerton, J. W., and Greenberg, E. P. 1998. The involvement of cell-to-cell signals in the development of a bacterial biofilm. Science **280**:295–298.

18. De Kievit, T. R., Gillis, R., Marx, S., Brown, C., and Iglewski, B. H. 2001. Quorum-sensing genes in *Pseudomonas aeruginosa* biofilms: Their role and expression patterns. Appl. Environ. Microbiol. **67**:1865–1873.

19. Drlica, K., and Perl-Rosenthal, N. R. 1999. DNA switches for thermal control of gene expression. Trends Microbiol. **7**:425–426.

20. Dunny, G. M., and Winans, S. C. 1999. *Cell–Cell Signaling in Bacteria*. ASM Press, Washington, D.C.
21. Freeman, J. A., and Bassler, B. L. 1999. A genetic analysis of the function of LuxO, a two-component response regulator involved in quorum sensing in *Vibrio harveyi*. Mol. Microbiol. **31**:665–677.
22. Freeman, J. A., Lilley, B. N., and Bassler, B. L. 2000. A genetic analysis of the functions of LuxN: A two-component hybrid sensor kinase that regulates quorum sensing in *Vibrio harveyi*. Mol. Microbiol. **35**:139–149.
23. Flavier, A. B., Clough, S. J., Schell, M. A., and Denny, T. P. 1997. Identification of 3-hydroxypalmitic acid methyl ester as a novel autoregulator controlling virulence in *Ralstonia solanacearum*. Mol. Microbiol. **26**:251–259.
24. Flynn, J. L., and Chan, J. 2001. Immunology of tuberculosis. Annu. Rev. Immunol. **19**:93–129.
25. Foster, T. J., and Hook, M. 1998. Surface protein adhesins of *Staphylococcus aureus*. Trend. Microbiol. **6**:484–488.
26. Fuqua, C., and Eberhard, A. 1999. Signal generation in autoinducer systems: Synthesis of acylated homoserine lactones by LuxI-type proteins. In *Cell–Cell Signaling in Bacteria* (edited by G. M. Dunny and S. C. Winans), pp. 211–230. ASM Press, Washington, D.C.
27. Fuqua, W. C., Winans, S. C., and Greenberg, E. P. 1994. Quorum sensing in bacteria – the LuxR-LuxI family of cell density-responsive transcriptional regulators. J. Bacteriol. **176**:269–275.
28. Garcia-Lara, J., Shang, L. H., and Rothfield, L. I. 1996. An extracellular factor regulates expression of *sdiA*, a transcriptional activator of cell division genes in *Escherichia coli*. J. Bacteriol. **178**:2742–2748.
29. Gillaspy, A. F., Hickmon, S. G., Skinner, R. A., Thomas, J. R., Nelson, C. L., and Smeltzer, M. S. 1995. Role of the accessory gene regulator (*agr*) in pathogenesis of staphylococcal osteomyelitis. Infect. Immun. **63**:3373–3380.
30. Gray, K. M., and Garey, J. R. 2001. The evolution of bacterial LuxI and LuxR quorum sensing regulators. Microbiology **147**:2379–2387.
31. Groisman, E. A. 2001. The pleiotropic two-component regulatory system PhoP-PhoQ. J. Bacteriol. **183**:1835–1842.
32. Hanzelka, B. L., Parsek, M. R., Val, D. L., Dunlap, P. V., Cronan, J. E., Jr., and Greenberg, E. P. 1999. Acylhomoserine lactone synthase activity of the *Vibrio fischeri* AinS protein. J. Bacteriol. **181**:5766–5770.
33. Hassett, D. J., Ma, J. F., Elkins, J. G., McDermott, T. R., Ochsner, U. A., West, S. E., Huang, C. T., Fredericks, J., Burnett, S., Stewart, P. S., McFeters, G., Passador, L., and Iglewski, B. H. 1999. Quorum sensing in *Pseudomonas aeruginosa* controls expression of catalase and superoxide dismutase genes

122

SIMON SWIFT

and mediates biofilm susceptibility to hydrogen peroxide. Mol. Microbiol. **34**:1082–1093.

34. Hoiby, N., Krogh Johansen, H., Moser, C., Song, Z., Ciofu, O., and Kharazmi, A. 2001. *Pseudomonas aeruginosa* and the *in vitro* and *in vivo* biofilm mode of growth. Microbes Infect. **3**:23–35.
35. Holden, M. T. G., Chhabra, S. R., de Nys, R., Stead, P., Bainton, N. J., Hill, P. J., Manefield, M., Kumar, N., Labatte, M., England, D., Rice, S., Givskov, M., Salmond, G. P. C., Stewart, G. S. A. B., Bycroft, B. W., Kjelleberg, S., and Williams, P. 1999. Quorum-sensing cross talk: Isolation and chemical characterization of cyclic dipeptides from *Pseudomonas aeruginosa* and other Gram-negative bacteria. Mol. Microbiol. **33**:1254–1266.
36. Holden, M., Swift, S., and Williams, P. 2000. New signal molecules on the quorum-sensing block. Trends Microbiol. **8**:101–104.
37. Hurme, R., and Rhen, M. 1998. Temperature sensing in bacterial gene regulation – what it all boils down to Mol. Microbiol. **30**:1–6.
38. Jakubovics, N. S., and Jenkinson, H. F. 2001. Out of the iron age: new insights into the critical role of manganese homeostasis in bacteria. Microbiology. **147**:1709–1718.
39. Ji, G., Beavis, R. C., and Novick, R. P. 1995. Cell density control of staphylococcal virulence mediated by an octapeptide pheromone. Proc. Natl. Acad. Sci. USA **92**:12,055–12,059.
40. Ji, G., Beavis, R., and Novick, R. P. 1997. Bacterial interference caused by autoinducing peptide variants. Science **276**:2027–2030.
41. Jiang, M., Grau, R., and Perego, M. 2000a. Differential processing of propeptide inhibitors of Rap phosphatases in *Bacillus subtilis*. J. Bacteriol. **182**:303–310.
42. Jiang, M., Shao, W., Perego, M., and Hoch, J. A. 2000b. Multiple histidine kinases regulate entry into stationary phase and sporulation in *Bacillus subtilis*. Mol. Microbiol. **38**:535–542.
43. Kaplan, H. B., and Greenberg, E. P. 1985. Diffusion of autoinducer is involved in regulation of the *Vibrio fischeri* luminescence system. J. Bacteriol. **163**:1210–1214.
44. Kaprelyants, A. S., and Kell, D. B. 1993. Dormancy in stationary-phase cultures of *Micrococcus luteus*: Flow cytometric analysis of starvation and resuscitation. Appl. Environ. Microbiol. **59**:3187–3196.
45. Kell, D. B., and Young, M. 2000. Bacterial dormancy and culturability: The role of autocrine growth factors. Curr. Opin. Microbiol. **3**:238–243.
46. Kleerebezem, M., Quadri, L. E., Kuipers, O. P., and de Vos, W. M. 1997. Quorum sensing by peptide pheromones and two-component signal-transduction systems in Gram-positive bacteria. Mol. Microbiol. **24**:895–904.

47. Kleerebezem, M., de Vos, W. M., and Kuipers, O. P. 1999. The lantibiotics nisin and subtilin act as extracellular regulators of their own biosynthesis. In *Cell–Cell Signaling in Bacteria* (edited by G. M. Dunny and S. C. Winans), pp. 159–174. ASM Press, Washington, D.C.

48. Kunst, F., Ogasawara, N., Moszer, I., Albertini, A. M., Alloni, G., Azevedo, V., Bertero, M. G., Bessieres, P., Bolotin, A., Borchert, S., Borriss, R., Boursier, L., Brans, A., Braun, M., Brignell, S. C., Bron, S., Brouillet, S., Bruschi, C. V., Caldwell, B., Capuano, V., Carter, N. M., Choi, S. K., Codani, J. J., Connerton, I. F., Danchin, A., et al. 1997. The complete genome sequence of the Gram-positive bacterium *Bacillus subtilis*. Nature **390**:249–256.

49. Laue, B. E., Jiang, Y., Chhabra, S. R., Jacob, S., Stewart, G. S. A. B., Hardman, A., Downie, J. A., O'Gara, F., and Williams, P. 2000. The biocontrol strain *Pseudomonas fluorescens* F113 produces the *Rhizobium* "small" bacteriocin *N*-(3-hydroxy-7-*cis*-tetradecenoyl)-homoserine lactone via HdtS, a putative novel *N*-acylhomoserine lactone synthase. Microbiology **146**:2469–2480.

50. Lawrence, R. N., Dunn, W. R., Bycroft, B., Camara, M., Chhabra, S. R., Williams, P., and Wilson, V. G. 1999. The *Pseudomonas aeruginosa* quorum-sensing signal molecule, *N*-(3-oxododecanoyl)-L-homoserine lactone, inhibits porcine arterial smooth muscle contraction. Br. J. Pharmacol. **128**:845–848.

51. Lazazzera, B. A. 2000. Quorum sensing and starvation: Signals for entry into stationary phase. Curr. Opin. Microbiol. **3**:177–182.

52. Lazazzera, B. A., Palmer, T., Quisel, J., and Grossman, A. D. 1999. Cell density control of gene expression and development in *Bacillus subtilis*. In *Cell–Cell Signaling in Bacteria* (edited by G. M. Dunny and S. C. Winans), pp. 27–46. ASM Press, Washington, D.C.

53. Lilley, B. N., and Bassler, B. L. 2000. Regulation of quorum sensing in *Vibrio harveyi* by LuxO and sigma-54. Mol. Microbiol. **36**:940–954.

54. Lina, G., Jarraud, S., Ji, G., Greenland, T., Pedraza, A., Etienne, J., Novick, R. P., and Vandenesch, F. 1998. Transmembrane topology and histidine protein kinase activity of AgrC, the *agr* signal receptor in *Staphylococcus aureus*. Mol. Microbiol. **28**:655–662.

55. Lowy, F. D. 1998. *Staphylococcus aureus* infections. N. Engl. J. Med. **339**:520–532.

56. Lyczak, J. B., Cannon, C. L., and Pier, G. B. 2000. Establishment of *Pseudomonas aeruginosa* infection: Lessons from a versatile opportunist. Microbes Infect. **2**:1051–1060.

57. Lyon, G. J., Mayville, P., Muir, T. W., and Novick, R. P. 2000. Rational design of a global inhibitor of the virulence response in *Staphylococcus aureus*, based

SIMON SWIFT

in part on localization of the site of inhibition to the receptor-histidine kinase, AgrC. Proc. Natl. Acad. Sci. USA **97**:13,330–13,335.

58. Magnuson, R., Solomon, J., and Grossman, A. D. 1994. Biochemical and genetic characterization of a competence pheromone from *B. subtilis.* Cell **77**:207–216.

59. Mayville, P., Ji, G., Beavis, R., Yang, H., Goger, M., Novick, R. P., and Muir, T. W. 1999. Structure-activity analysis of synthetic autoinducing thiolactone peptides from *Staphylococcus aureus* responsible for virulence. Proc. Natl. Acad. Sci. USA **96**:1218–1223.

60. McDowell, P., Affas, Z., Reynolds, C., Holden, M. T., Wood, S. J., Saint, S., Cockayne, A., Hill, P. J., Dodd, C. E. R., Bycroft, B. W., Chan, W. C., and Williams, P. 2001. Structure, activity and evolution of the group I thiolactone peptide quorum-sensing system of *Staphylococcus aureus*. Mol. Microbiol. **41**:503–512.

61. Michael, B., Smith, J. N., Swift, S., Heffron, F., and Ahmer, B. M. M. 2001. SdiA of *Salmonella enterica* is a LuxR homolog that detects mixed microbial communities. J. Bacteriol. **183**:5733–5742.

62. Michiels, J., Dirix, G., Vanderleyden, J., and Xi, C. 2001. Processing and export of peptide pheromones and bacteriocins in Gram-negative bacteria. Trends Microbiol. **9**:164–168.

63. Miller, M. B., and Bassler, B. L. 2001. Quorum sensing in bacteria. Annu. Rev. Microbiol. **55**:165–199.

64. Milton, D. L., Chalker, V. J., Kirke, D., Hardman, A., Camara, M., and Williams, P. 2001. The LuxM homologue VanM from *Vibrio anguillarum* directs the synthesis of *N*-(3-hydroxyhexanoyl)homoserine lactone and *N*-hexanoylhomoserine lactone. J. Bacteriol. **183**:3537–3547.

65. Mueller, J. P., Bukusoglu, G., and Sonenshein, A. L. 1992. Transcriptional regulation of *Bacillus subtilis* glucose starvation-inducible genes: Control of *gsiA* by the ComP-ComA signal transduction system. J. Bacteriol. **174**:4361–4373.

66. Mukamolova, G. V., Kaprelyants, A. S., Young, D. I., Young, M., and Kell, D. B. 1998. A bacterial cytokine. Proc. Natl. Acad. Sci. USA **95**:8916–8921.

67. Navarre, W. W., and Schneewind, O. 1999. Surface proteins of gram-positive bacteria and mechanisms of their targeting to the cell wall envelope. Microbiol. Mol. Biol. Rev. **63**:174–229.

68. Nicholson, W. L., Munakata, N., Horneck, G., Melosh, H. J., and Setlow, P. 2000. Resistance of *Bacillus* endospores to extreme terrestrial and extraterrestrial environments. Microbiol. Mol. Biol. Rev. **64**:548–572.

69. Nölling, J., Breton, G., Omelchenko, M. V., Makarova, K. S., Zeng, Q., Gibson, R., Lee, H. M., Dubois, J., Qiu, D., Hitti, J., Wolf, Y. I., Tatusov, R. L., Sabathe, F., Doucette-Stamm, L., Soucaille, P., Daly, M. J., Bennett, G. N., Koonin, E. V., and Smith, D. R. 2001. Genome sequence and comparative analysis of the solvent-producing bacterium. *Clostridium acetobutylicum*. J. Bacteriol. **183**:4823–4838.

70. Novick, R. P. 1999. Regulation of pathogenicity in *Staphylococcus aureus* by a peptide-based density-sensing system. In *Cell–Cell Signaling in Bacteria* (edited by G. M. Dunny and S. C. Winans), pp. 129–146. ASM Press, Washington, D.C.

71. Novick, R. P., Ross, H. F., Projan, S. J., Kornblum, J., Kreiswirth, B., and Moghazeh, S. 1993. Synthesis of staphylococcal virulence factors is controlled by a regulatory RNA molecule. EMBO J. **12**:3967–3975.

72. Novick, R. P., Projan, S. J., Kornblum, J., Ross, H. F., Ji, G., Kreiswirth, B., Vandenesch, F., and Moghazeh, S. 1995. The *agr* P2 operon: An autocatalytic sensory transduction system in *Staphylococcus aureus*. Mol. Gen. Genet. **248**:446–458.

73. Parsek, M. R., and Greenberg, E. P. 2000. Acyl-homoserine lactone quorum sensing in Gram-negative bacteria: A signaling mechanism involved in associations with higher organisms. Proc. Natl. Acad. Sci. USA **97**:8789–8793.

74. Pearson, J. P., Van Delden, C., and Iglewski, B. H. 1999. Active efflux and diffusion are involved in transport of *Pseudomonas aeruginosa* cell-to-cell signals. J. Bacteriol. **181**:1203–1210.

75. Pearson, J. P., Feldman, M., Iglewski, B. H., and Prince, A. 2000. *Pseudomonas aeruginosa* cell-to-cell signaling is required for virulence in a model of acute pulmonary infection. Infect. Immun. **68**:4331–4334.

76. Peng, H. L., Novick, R. P., Kreiswirth, B., Kornblum, J., and Schlievert, P. 1988. Cloning, characterization, and sequencing of an accessory gene regulator (*agr*) in *Staphylococcus aureus*. J. Bacteriol. **170**:4365–4372.

77. Perego, M. 1997. A peptide export-import control circuit modulating bacterial development regulates protein phosphatases of the phosphorelay. Proc. Natl. Acad. Sci. USA **94**:8612–8617.

78. Perego, M. 1998. Kinase-phosphatase competition regulates *Bacillus subtilis* development. Trends Microbiol. **6**:366–370.

79. Perego, M., 1999. Self-signaling by Phr peptides modulates *Bacillus subtilis* development. In *Cell–Cell Signaling in Bacteria* (edited by G. M. Dunny and S. C. Winans), pp. 243–258. ASM Press, Washington, D.C.

80. Perego, M., Higgins, C. F., Pearce, S. R., Gallagher, M. P., and Hoch, J. A. 1991. The oligopeptide transport system of *Bacillus subtilis* plays a role in the initiation of sporulation. Mol. Microbiol. **5**:173–185.

81. Perego, M., Hanstein, C., Welsh, K. M., Djavakhishvili, T., Glaser, P., and Hoch, J. A. 1994. Multiple protein-aspartate phosphatases provide a mechanism for the integration of diverse signals in the control of development in *B. subtilis*. Cell **79**:1047–1055.

82. Perego, M., Glaser, P., and Hoch, J. A. 1996. Aspartyl-phosphate phosphatases deactivate the response regulator components of the sporulation signal transduction system in *Bacillus subtilis*. Mol. Microbiol. **19**:1151–1157.

83. Pesci, E. C., Milbank, J. B., Pearson, J. P., McKnight, S., Kende, A. S., Greenberg, E. P., and Iglewski, B. H. 1999. Quinolone signaling in the cell-to-cell communication system of *Pseudomonas aeruginosa*. Proc. Natl. Acad. Sci. USA **96**:11,229–11,234.

84. Predich, M., Nair, G., Smith, I. 1992. *Bacillus subtilis* early sporulation genes *kinA*, *spoOF*, and *spoOA* are transcribed by the RNA polymerase containing sigma H. J. Bacteriol. **174**:2771–2778.

85. Qin, Y., Luo, Z. Q., Smyth, A. J., Gao, P., Beck von Bodman, S., and Farrand, S. K. 2000. Quorum-sensing signal binding results in dimerization of TraR and its release from membranes into the cytoplasm. EMBO J. **19**:5212–5221.

86. Ratledge, C., and Dover, L. G. 2000. Iron metabolism in pathogenic bacteria. Annu. Rev. Microbiol. **54**:881–941.

87. Raupach, B., and Kaufmann, S. H. 2001. Immune responses to intracellular bacteria. Curr. Opin. Immunol. **13**:417–428.

88. Rudner, D. Z., LeDeaux, J. R., Ireton, K., and Grossman, A. D. 1991. The *spoOK* locus of *Bacillus subtilis* is homologous to the oligopeptide permease locus and is required for sporulation and competence. J. Bacteriol. **173**:1388–1398.

89. Rumbaugh, K. P., Griswold, J. A., Iglewski, B. H., and Hamood, A. N. 1999. Contribution of quorum sensing to the virulence of *Pseudomonas aeruginosa* in burn wound infections. Infect. Immun. **67**:5854–5862.

90. Rumbaugh, K. P., Griswold, J. A., and Hamood, A. N. 2000. The role of quorum sensing in the in vivo virulence of *Pseudomonas aeruginosa*. Microbes Infect. **2**:1721–1731.

91. Schauder, S., and Bassler, B. L. 2001. The languages of bacteria. Genes Dev. **15**:1468–1480.

92. Schauder, S., Shokat, K., Surette, M. G., and Bassler, B. L. 2001. The LuxS family of bacterial autoinducers: Biosynthesis of a novel quorum-sensing signal molecule. Mol. Microbiol. **41**:463–476.

93. Shinefield, H. R., Ribble, J. C., and Boris, M. 1971. Bacterial interference between strains of *Staphylococcus aureus*, 1960 to 1970. Am. J. Dis. Child. **121**:148–152.

INTERBACTERIAL SIGNALLING IN DORMANCY AND DISEASE

94. Simonen, M., and Palva, I. 1993. Protein secretion in *Bacillus* species. Microbiol. Rev. **57**:109–137.
95. Sitnikov, D. M., Schineller, J. B., and Baldwin, T. O. 1996. Control of cell division in *Escherichia coli*: Regulation of transcription of *ftsQA* involves both *rpoS* and SdiA-mediated autoinduction. Proc. Natl. Acad. Sci. USA **93**:336–341.
96. Sonenshein, A. L. 2000. Control of sporulation initiation in *Bacillus subtilis*. Curr. Opin. Microbiol. **3**:561–566.
97. Swift, S., Throup, J. P., Williams, P., Salmond, G. P. C., and Stewart, G. S. A. B. 1996. Quorum sensing: A population-density component in the determination of bacterial phenotype. Trends Biochem. Sci. **21**:214–219.
98. Swift, S., Downie, J. A., Whitehead, N. A., Barnard, A. M. L., Salmond, G. P. C., and Williams, P. 2001. Quorum sensing as a population density dependent determinant of bacterial physiology. Adv. Microb. Physiol. **45**:199–270.
99. Tang, H. B., DiMango, E., Bryan, R., Gambello, M., Iglewski, B. H., Goldberg, J. B., and Prince, A. 1996. Contribution of specific *Pseudomonas aeruginosa* virulence factors to pathogenesis of pneumonia in a neonatal mouse model of infection. Infect. Immun. **64**:37–43.
100. Telford, G., Wheeler, D., Williams, P., Tomkins, P. T., Appleby, P., Sewell, H., Stewart, G. S. A. B., Bycroft, B. W., and Pritchard, D. I. 1998. The *Pseudomonas aeruginosa* quorum-sensing signal molecule *N*-(3-oxododecanoyl)-L-homoserine lactone has immunomodulatory activity. Infect. Immun. **66**:36–42.
101. Van Delden, C., and Iglewski, B. H. 1998. Cell-to-cell signaling and *Pseudomonas aeruginosa* infections. Emerg. Infect. Dis. **4**:551–560.
102. Van Delden, C., Pesci, E. C., Pearson, J. P., and Iglewski, B. H. 1998. Starvation selection restores elastase and rhamnolipid production in a *Pseudomonas aeruginosa* quorum-sensing mutant. Infect. Immun. **66**:4499–4502.
103. Vuong, C., Saenz, H. L., Gotz, F., and Otto, M. 2000. Impact of the *agr* quorum-sensing system on adherence to polystyrene in *Staphylococcus aureus*. J. Infect. Dis. **182**:1688–1693.
104. Wang, X. D., de Boer, P. A., and Rothfield, L. I. 1991. A factor that positively regulates cell division by activating transcription of the major cluster of essential cell division genes of *Escherichia coli*. EMBO J. **10**:3363–3372.
105. Wang, D., Ding, X., and Rather, P. N. 2001. Indole can act as an extracellular signal in *Escherichia coli*. J. Bacteriol. **183**:4210–4216.
106. Waterer, G. W., and Wunderink, R. G. 2001. Increasing threat of Gram-negative bacteria. Crit. Care Med. **29**:N75–81.

SIMON SWIFT

107. Weir, J., Predich, M., Dubnau, E., Nair, G., and Smith, I. 1991. Regulation of *spo0H*, a gene coding for the *Bacillus subtilis* sigma H factor. J. Bacteriol. **173**:521–529.

108. Whitehead, N. A., Barnard, A. M. L., Slater, H., Simpson, N. J. L., and Salmond, G. P. C. 2001. Quorum-sensing in Gram-negative bacteria. FEMS Microbiol. Rev. **25**:365–404.

109. Whiteley, M., Lee, K. M., and Greenberg, E. P. 1999. Identification of genes controlled by quorum sensing in *Pseudomonas aeruginosa*. Proc. Natl. Acad. Sci. USA **96**:13,904–13,909.

110. Winzer, K., and Williams, P. 2001. Quorum sensing and the regulation of virulence gene expression in pathogenic bacteria. Int. J. Med. Microbiol. **291**:131–143.

111. Withers, H. L., and Bernander, R. 1998. Characterization of *dnaC2* and *dnaC28* mutants by flow cytometry. J. Bacteriol. **180**:1624–1631.

112. Withers, H. L., and Nordström, K. 1998. Quorum-sensing acts at initiation of chromosomal replication in *Escherichia coli*. Proc. Natl. Acad. Sci. USA **95**:15,694–15,699.

113. Withers, H., Swift, S., and Williams, P. 2001. Quorum-sensing as an integral component of gene regulatory networks in Gram-negative bacteria. Curr. Opin. Microbiol. **4**:186–193.

114. World Health Organization. 2001. *Global Tuberculosis Control. WHO Report 2001.* WHO, Geneva.

115. Wood, J. M. 1999. Osmosensing by bacteria: Signals and membrane-based sensors. Microbiol. Mol. Biol. Rev. **63**:230–262.

CHAPTER 5

Mechanisms of stationary-phase mutagenesis in bacteria and their relevance to antibiotic resistance

Digby F. Warner and Valerie Mizrahi

(131)

INTRODUCTION

Studies of spontaneous mutation rates in bacteria have tended to utilise exponentially growing populations, despite the fact that in their natural environments cells probably spend only a fraction of their lives in log-phase growth (85). Of equal relevance, especially clinically, are questions pertaining to the physiological state of bacteria during latent disease, and the effects of an altered physiology on mutagenesis. A common misconception is that spontaneous mutations arise as a consequence of errors made exclusively during DNA replication in actively growing cells. As early as the 1950s, Ryan (116) showed that mutations could arise in apparently static (or non-dividing) bacterial populations subjected to non-lethal selective conditions. Subsequent work extending this observation has redefined our understanding of the fundamental environmental and intracellular conditions promoting mutagenesis (18). While actively growing cells have the advantage of elevated rates of mutation per unit time, the increased periods spent in stationary phase by most cellular populations mean that even very low mutation rates can lead to the accumulation of high mutation frequencies per cell (13). Compounding the effects of an extended stationary phase are inducible systems, such as the bacterial SOS response (139) and its associated mutator polymerases (37), as well as hypermutable genes (29), all of which have potentially important implications for the development of subpopulations of bacteria with genetically encoded drug and stress resistance.

THE STATIONARY PHASE

The physiological state adopted by many bacterial pathogens that allows extended survival within the host, sometimes in the absence of associated disease (e.g., latent *Mycobacterium tuberculosis* infection), is poorly understood. Although clinically latent bacteria are often referred to as dormant, in its strictest sense this term should be reserved for a "reversible state of low metabolic activity, in which cells can persist for extended periods without cell division" (56). Emergence from dormancy generally requires specific molecular signals, in contrast to stationary-phase exit, which is induced by a favourable change in environmental conditions. The resuscitation factors mediating recovery of dormant bacteria are distinguished from nutrients in that they are produced by the organisms themselves, are active at very low concentrations and, unless generated from pro-hormones, do not require metabolism for activity (88). Most organisms do not enter a dormant state as defined and represented in nutrient-starved *Micrococcus luteus* (54), nor do they adopt a physiological state analogous to that observed in *Bacillus* species following sporulation (27). Therefore, clinically latent infections are probably composed of heterogeneous populations of metabolically active bacilli in a state of non-growth or prolonged stationary phase; in such populations, despite very slow growth and limited cell turnover, the net viable cell count remains constant (97).

Signals for Entry into Stationary Phase

Entry of bacteria into stationary phase is characterised by a dramatic alteration in gene expression profiles to allow extended survival in the face of metabolic and genotoxic stress. In a sense, stationary-phase cells may be considered the functional equivalents of spores (136), and the stationary phase itself can be considered both a response to environmental conditions and a differentiated physiological state (119). However, the heterogeneity of stationary-phase populations should be borne in mind, along with the fact that the physiology of the cells within those populations is determined by the conditions that prompted the cessation of growth (51). The primary cue for entry into stationary phase is probably the absence of sufficient nutrients to sustain growth (59). However, a host of factors other than nutrient starvation have been implicated in stationary-phase regulation. These include the accumulation of toxic metabolites, and environmental stresses such as extremes in temperature or pH, low oxygen tension, high osmolarity, and oxidative agents (52, 120, 145). There is also evidence that quorum sensing regulates the transition into stationary phase in some bacterial species, possibly in conjunction with starvation signals mediated by key transcription factors (66). *In vivo*, the hostile environments encountered by commensal and pathogenic bacteria – as an inevitable consequence of their specific locus of habitation, or as a result of host immune surveillance and detection, or the action of antibiotic compounds – probably provide strong signals for entry into stationary phase.

Morphological and Physiological Manifestations

In some bacteria, the stationary phase manifests morphologically in the form of specialised structures, such as the highly resistant endospores described in *Bacillus subtilis* (27), or the fruiting bodies observed in *Myxococcus* species (57). However, many bacterial species do not form specialised structures, and the morphological and physiological changes that occur in such non-differentiating bacteria include a decrease in cell size, altered cell wall structure, increase in periplasmic volume, cytoplasmic condensation, increased stress resistance, altered protein synthetic profiles, increased RNA stability, and reductive cell division (51, 61). Natural genetic competence may be induced upon entry into stationary phase or by growth inhibition in bacteria such as *Haemophilus influenzae* (24), while in *Bacillus subtilis* the same factors involved in initiation of sporulation can promote the adoption of a transformation competent state (41). Apart from the increased potential for acquiring antibiotic or other stress resistance elements, natural competence has been shown in *Escherichia coli* to alleviate starvation by allowing uptake of exogenous DNA as an alternative carbon and energy source (31).

Studies on the starvation-induced stationary-phase kinetics of several non-sporulating bacterial species have shown that long-term survival is dependent on continued differential protein synthesis (54, 72, 92, 140) and that proteins are temporally expressed throughout the different stages of starvation (65, 105, 140). The altered protein synthetic profiles seen in these starved cultures, and the transcriptional variation they demand, provide molecular biological evidence of an active stationary-phase metabolism and further contradict the idea of a quiescent stationary-phase population.

DNA Damage and Metabolism in the Stationary Phase

Although the DNA of resting cells is well protected (51), it may be subjected to insults from both exogenous and endogenous sources, depending on the particular environment encountered by the organism. Several forms of endogenous DNA damage have been implicated in the emergence of adaptive mutations, and spontaneous DNA damage has been demonstrated repeatedly in apparently static cells (43). Nitroso compounds, for example, are major contributors of alkylation damage (130), while the repair of oxidative damage in stationary-phase cells has been linked to adaptive mutagenesis

(6). As expected, increased mutation rates have been demonstrated in strains defective in the repair of these types of DNA lesions. Adaptive reversion of a *his* mutation is increased in the absence of *ogt* and *ada* (104), confirming the role of alkylation damage in static cells. Similarly, oxidative damage to DNA results in increased rates of adaptive reversion in resting *mutY* and *mutM mutY* strains (11, 14). A lesion in *mutT* dramatically increases adaptive reversion (12), providing further evidence that mutations arise during DNA synthesis in stationary cells. However, the massive discrepancy between the observed increase in reversion in the *mutT* mutant and that predicted from an estimate of DNA turnover in resting cells (32) is suggestive either of an increased polymerase error rate in static cells or of an amount of DNA turnover not detectable by the methods employed.

Mutations in Stationary Phase Cells

While several bacterial and yeast assay systems have demonstrated stationary-phase mutations in cells exposed to a variety of exogenous stresses (35, 53, 124), few common underlying principles have been identified. Instead, the main conclusion from these studies is that the mutagenic mechanism induced in each is a function of the specific environmental stress applied and the genotypes of the cells assayed. That is, stationary-phase mutation describes a collection of cellular responses to non-lethal stresses in which mutations are promoted (74). Among the mutation mechanisms reported are transposition events (82, 89), substitution and frameshift mutations (146, 149), and gene amplifications (122). Of course, only some of the induced mutations confer a growth advantage in the limiting environment and so are selected, whereas the majority confer no apparent advantage. Nevertheless, it is the small subpopulation of advantageous mutations that has important implications for stationary-phase survival.

Several studies have demonstrated the presence of dynamic, heterogeneous populations of bacteria in the stationary phase. Starved cultures of *E. coli* undergo successive rounds of population takeover by mutants of increased fitness expressing a growth advantage in stationary phase (GASP) phenotype, while mutations in genes involved in amino acid catabolism or respiratory pathways affect stationary-phase survival (144–146, 149). Similarly, during prolonged stationary phase, variants of *Mycobacterium smegmatis* arise that out-compete exponential phase-adapted strains in stationary-phase culture (120). In addition, the mutation rate appears to vary with altered growth conditions, and starvation has often been implicated in mutagenesis

(50). Starved *E. coli* cells show an increase in mutation to rifampin resistance mediated by cAMP-dependent SOS induction (125) and, possibly, DNA polymerase (pol) IV–mediated mutagenesis. Furthermore, a decrease in the levels of MutH and MutS mismatch repair (MMR) proteins has been observed in stationary-phase *E. coli*, implying an impaired repair capacity in starved cells (28). In *M. smegmatis*, mutation rates seem to increase significantly during the post-exponential growth phase (55). However, error avoidance enzymes may also be upregulated in stationary-phase cells (84).

ADAPTIVE EVOLUTION OF MUTATION RATES

Molecular biological evidence has shown that several genetic factors determine the rate and spectrum of mutations in an organism. The fact that these genetic factors are themselves subject to adaptive evolution raises an intriguing question: is it possible for organisms to modulate mutational dynamics under variable selective conditions? It has been suggested that in bacteria, the SOS response and hypervariable loci may have evolved by second-order selection to provide just such mutational modulation while limiting the cost associated with adaptive mutagenesis (79, 103, 132). As Metzgar and Wills (79) have cogently argued, mutational tuning need not violate fundamental evolutionary principles. Mutations are simply increased under certain environmental conditions or targeted to specific regions of the genome: in other words non-randomness with respect to function, as the specific generation of adaptive mutations implies, is replaced by non-randomness with respect to time or location.

A distinction should be made between heritable and environment-dependent mechanisms of mutation rate modification (79). Heritable mutator mechanisms are independent of the environment and alter the global mutation rate or target mutations to hotspots of mutation. The mutator phenotypes of mutator mutants are therefore constitutive and induce mutations irrespective of the applied environmental conditions. By contrast, inducible mutators act in a genome-wide (global) manner to increase mutation rates specifically in response to applied environmental conditions. Inducible mutagenesis may therefore be an adaptive response that has been selected for, whereas the acquisition of a mutator phenotype is a random event. Of course, mutator mutants may exist in tandem with inducible mutators in stationary phase, and a heterogeneous population, consisting of transient and heritable mutators, might be the best equipped to respond rapidly to the bottlenecks encountered by most pathogens *in vivo*.

Heritable Global Mutators

Mutations in DNA replication or repair genes involved in mutation avoidance may confer a mutator phenotype in terms of genome stability, strain variability, and mutation to drug or stress resistance (49). Mutator mutants comprise a small proportion of both pathogenic and commensal *E. coli* and *Salmonella typhimurium* strains (71), although strong mutators are associated primarily with pathogenic bacteria and exhibit an MMR-defective phenotype (67). Recently, *mutY* mutator mutants were isolated from *E. coli* populations under nutrient starvation *in vitro* (91), mirroring their previous identification in clinical strains of *Pseudomonas aeruginosa* isolated from cystic fibrosis patients (94). The presence of a small proportion of heritable mutators among the bacterial population in a chronic infection suggests that an elevated mutation rate may allow pathogenic bacteria to adapt better to the fluctuating environment within the host (127). However, the risks of deleterious mutations are elevated in mutators, and the limited number of mutator mutants may indicate that the benefits associated with a high mutation rate are transient and that regaining the wild-type genotype is essential to the long-term survival of the population (39, 40).

Heritable Local Mutators

Specific regions of a genome may be predisposed to mutation by virtue of their unique sequence characteristics. Tandem repeats (microsatellites) perturb DNA polymerase activity resulting in replication slippage errors by single-motif insertion and deletion and so constitute hotspots of mutation. Although they are up to 10^4 times more likely to occur than base substitution errors, polymerase slippage errors are very efficiently corrected by MMR (64). However, even in wild-type cells, microsatellite sequences render genes susceptible to frameshift mutagenesis, and this susceptibility is greatly increased in MMR-deficient mutators. "Contingency genes," which carry hotspots of mutation within the ORF or in its regulatory elements, typically encode strongly selected structures such as surface antigens or restriction/modification systems in pathogenic bacterial species (86). The association of hypermutable loci with genes encoding important biological functions that are subject to intense selective pressure suggests that these sequences have not evolved by chance (86). This observation is supported by the heterogeneity of the repeat DNA sequences driving hypermutability, as well as the diverse nature of the proteins encoded by contingency genes (29, 87). Furthermore, the fact that most contingency repeats constitute the longest microsatellite sequences in their respective prokaryotic genomes suggests that repetition itself may be under selection (79).

The elevated mutation rates that microsatellites confer on contingency genes, and the associated increase in allelic variation, improve the chances of host immune response evasion. In such circumstances, where selection for diversity is restricted to only a few genes, a localised increased mutation rate may be selected for. The advantage of this evolutionary strategy is that the localised mutator avoids the costs associated with generalised mutagenesis. However, the evolution of adaptively variable antigenic determinants in pathogens imposes additional selective pressure on the host to evolve similarly adaptive immune defences. The adaptive dynamic that is thus established between host and pathogen creates an optimal environment for the evolution and possibly indefinite escalation of mutators (44).

Environment-Dependent Mutators

Inducible mutators operate in organisms maladapted to the prevailing conditions. By definition, the mutator phenotype is induced in response to applied environmental conditions and so might be considered more relevant to the emergence of adaptive mutants in a population of bacteria under stress. Those cells that survive, either by genetic adaptation or through a favourable environmental change, produce progeny cells with normal mutation rates. In this way, the offspring are not burdened with the risk of increased rates of deleterious mutation. A transient increase in mutation rate is reminiscent of the adaptive mutations produced under starvation conditions in *E. coli*. In fact, the bacterial SOS response may be considered the prototypical environment-dependent mutator system. Although not as well characterised as the SOS system, the identification of an additional environment-dependent mutator system, ROSE (125, 126), reinforces the potential evolutionary importance of mutagenic machinery under cellular control and capable of transient adjustment to the mutation rate in response to imposed conditions.

Error-Prone DNA Polymerases

The mutator polymerases involved in SOS mutagenesis form part of a recently discovered DNA polymerase superfamily comprising the widely distributed DinB, UmuC, Rad30, and Rev1 subfamilies (37, 38). Many members of this superfamily function in DNA damage tolerance or repair, and some have been shown to make errors >100-fold more frequently than high fidelity, constitutive polymerases (93, 128, 129, 137, 147). In fact, the error-prone nature of these enzymes provided the first clue that they might be involved in mutational processes under cellular control (102). Mutator polymerases with known lesion-bypass activity such as *E. coli* pol V (UmuD$'_2$C) (128, 129) are believed to replace the highly processive replication machinery when the latter becomes stalled (37). Since they are liable to insert an incorrect nucleotide, mutator polymerases probably promote mutations during replicative bypass (99). After lesion bypass, the mutator polymerase is displaced, and the original replication machinery returns to highly processive, high fidelity synthesis. The *E. coli dinB* gene[1] encoding pol IV is the first in an operon of four damage-inducible SOS genes (20). Pol IV has no proofreading activity, is strictly distributive, and promotes base deletions and single-nucleotide substitutions at undamaged template sites (58, 137, 138). The fact that untargeted mutations often appear as genuine base-pairing errors initially implicated pol IV as a possible genome-wide mutator polymerase acting under a variety of metabolic stresses. The mutagenic properties of pol IV are well documented: cells carrying a *dinB* null allele are defective in λ untargeted mutagenesis (15, 58), whereas over-production of pol IV leads to a significant increase in mutation in the absence of DNA-damaging agents (58, 138). Both substitution and frameshift mutation frequencies are elevated, with frameshifts targeted predominantly at mononucleotide repeats. Although pol IV appears to function in DNA synthesis at misaligned replication forks, initial data suggest that it is not involved in trans-lesion synthesis at sites of damaged bases (128).

ADAPTIVE MUTAGENESIS

In broad terms, adaptive mutations are those that allow cells to overcome natural or artificial selective pressures. However, since its use by Delbrück (22) to describe mutations formed specifically in response to the environment in which they were selected, the term "adaptive mutation" has become associated with a particular set of mutations and mutational mechanisms occurring in well defined assay systems after exposure to a non-lethal selective medium. In these systems, the definition of adaptive mutation has been progressively refined and updated to explain and incorporate the latest empirical evidence. Key to the concept of adaptive mutagenesis is the requirement that adaptive mutations arise solely in response to the environment in

---

[1] Although *dinP* is often used in sequence annotations, the designation *dinB* has precedence and is used throughout this chapter.

which they are selected and so must be distinguished from mutations occurring during non-selected growth (35, 114). Furthermore, although utility is inherent in the term "adaptive," it should not be inferred that other, non-adaptive (or non-selected) mutations do not occur during selection, or that the adaptive mutations are induced preferentially. Adaptive mutation, its history and development, have been extensively reviewed (35, 114) and the reader is referred to these articles for detailed information on this topic. In this chapter, we provide a brief historical perspective and then describe those aspects of adaptive mutation that have specific relevance to stationary-phase mutagenesis in clinically latent bacterial populations.

Historical Perspective

In their seminal paper, Cairns *et al.* (18) stated "populations of bacteria, in stationary phase, have some way of producing (or selectively retaining) only the most appropriate mutations." This observation, though controversial and subsequently disproved, launched a new field of research and led to a redefinition of the role of mutation in evolution. The conclusion by Cairns *et al.* (18) was based on the observed genetic adaptation of *E. coli* to prolonged non-lethal selection. However, the claim that only adaptive (advantageous) mutants seemed to arise, and that mutagenesis *required* the selective pressure and so was "directed," had strong overtones of Lamarckism and placed the phenomenon squarely in opposition to Darwinian evolution. A fundamental tenet of Darwinian theory holds that mutation is a random event, occurring without prescience; in fact, evolution may be considered the interplay between the generation of random genetic variants and their phenotypic selection (103). In Darwinian terms, natural selection functions as a post-mutational sorting mechanism (79). Central to this dynamic is the principle that the event promoting a mutation is independent of the selective force; neo-Darwinists argue further that the rate of variation (mutation) is constant, thereby implying that it is natural selection, and not the creation of genetic variants, that drives evolution.

However, prior to these observations, Radman (101) and Echols (25) had suggested that states of accelerated evolution might be induced in response to stress and had predicted the existence of specialised enzymes that could hasten evolution under conditions in which genetic stasis might prove detrimental to survival. Since various environmental and nutrient stresses may prompt entry into stationary phase, the concept of increased mutation rates in such circumstances contradicted the belief that spontaneous mutations arise solely from DNA replication errors. In fact, Ryan had questioned this contention in the 1950s (117): confronted with the evidence of late-appearing His$^+$ revertants, he concluded that non-dividing cells must perform limited DNA synthesis and that this synthesis might be characterised by an increased error rate. Therefore, aside from its obvious Lamarckian implications, the proposal that stationary-phase bacteria were able to selectively mutate their genomes also challenged the idea of a quiescent stationary phase.

Adaptive Mutation in the E. coli FC40 Model System

The model system for studying adaptive mutagenesis and that from which most data have been obtained utilises a frameshift reversion assay in E. coli (35, 114). Strain FC40 is deleted for the lac operon in its chromosome, but carries a lac+1 frameshift allele on a low-copy F' episome. The strain cannot grow when plated on minimal medium containing lactose as sole carbon source (Lac$^-$), but growth-phase dependent Lac$^+$ revertants appear early, within two days of plating. Additional revertants to lactose utilisation accumulate over the next week and are distinguished from the growth-phase mutants in that they result from stationary-phase mutations occurring *after* exposure to the lactose medium (75). These late-arising, adaptive mutants are produced by two separate and fundamentally distinct processes, namely point mutation and gene amplification.

ADAPTIVE POINT MUTATIONS

Adaptive point mutation in FC40 is dependent on the homologous re-combination and double strand break repair (DSBR) proteins RecA, RecBC, and RuvABC (13). DSBR is thought to promote mutations in *cis* by initiating replication at double-stranded breaks (DSBs) or ends (45), after which the pol IV mutator polymerase introduces errors (74). This process therefore requires a functional SOS response and induced pol IV activity (74, 76). Whereas growth-dependent reversions are heterogeneous, adaptive point mutations are primarily characterised by −1 deletions in various small mononucleotide repeat sequences (36, 112), which resembles the spectrum of errors produced by pol IV *in vitro* (137). Simple repeat instability is associated with a lack of post-replicative MMR (83), and it has been shown that the adaptive mutations accumulate during a transient period of MMR protein deficiency during starvation (47). However, the data implicating MMR deficiency in adaptive point mutation are controversial. Although limiting MutL protein has mainly been implicated (46, but see 34), MutS might become limiting under certain environmental conditions (10, 148). Strangely, there is no decline in the number of MutL molecules per cell during stationary phase (46), which has prompted the suggestion that errors arising upon SOS-mediated induction of mutator DNA polymerases could titrate MMR and reduce its activity transiently (74).

Transient Hypermutability of a Sub population

In the FC40 system, an unusually high frequency of unrelated (unselected) mutations is observed in the genomes of the Lac$^+$ revertants, but not among similarly starved Lac$^-$ cells, which led to the suggestion that some or all of the adaptive mutants arise during a period of transient genome-wide hypermutability within a subpopulation (16, 111, 135). This view is in partial agreement with an earlier proposal (42) predicting a hypermutable state in a small proportion of cells under selective pressure. Hall, however, proposed that the subpopulation of hypermutable cells would either acquire the adaptive mutation or die as a result of error catastrophe (42). Contrary to this hypothesis, under non-lethal selection in the *lac* system there is an accumulation of non-selected stationary-phase mutations both on the F' episome (33) and on the chromosome (17) of Lac$^-$ cells, also generated by a recombination-dependent and pol IV-mediated mechanism. Evidence that the SOS response controls mutagenesis under starvation conditions, as demonstrated in both the *lac* (76) and ROSE (resting organisms in structured environments) systems (126), has raised the possibility that SOS induction might differentiate the two populations. However, the origin of the transiently hypermutable subpopulation is yet to be determined. Finally, mutations occur unevenly in Lac$^+$ adaptive mutants (135), indicating the presence of possible hypermutation hotspots, and it has been suggested that mutational "hot" and "cold" regions in the chromosome may be determined by their proximity to sites at which DSBs occur (113).

Extrapolation to Other Systems

Reversion to Lac$^+$ in FC40 has been so extensively studied that it has become the paradigm of adaptive mutation. Although not all aspects of the phenomenon as it manifests in this strain can be extrapolated to other systems, several valuable insights have been gained into some of the genetic and environmental factors required for adaptation to non-lethal stress. Such insights are potentially important to an understanding of the dynamics of microbial populations under selective pressure in their natural environments, in particular pathogenic bacteria occupying specialised niches within a host; but as with any model system, there are dangers inherent in using *in vitro* data to explain or predict *in vivo* outcomes and in generalising observations made in this system to others.

ADAPTIVE GENE AMPLIFICATION

Gene amplification has been identified as a fundamental evolutionary response in bacterial populations exposed to various selective pressures (108, 110, 123) and has been implicated, for instance, in adaptation to heavy-metal toxicity (62) and to growth on limiting carbon sources (122, 134). The principal benefit of gene amplification is that it is freely reversible by homologous recombination, allowing adjustments to the genome while retaining an unaltered copy of the original gene (95). Importantly, amplification generally does not lead to loss of gene function (2, 115) but enables over-expression of defined genomic regions that may be required for short-term adaptation. The notion that gene amplification might allow dynamic adaptation to applied environmental conditions demands that the amplified state persist only in the presence of the selective condition and that removal of the selective pressure result in haploid segregants dominating the bacterial population. Several natural instances relating to the acquisition of antibiotic resistance by various bacteria fulfil these requirements (26, 73, 90, 98, 143). In these cases, the increase in antibiotic resistance correlates with the degree of gene amplification, and subsequent culturing in antibiotic-free media results in deamplification and a concomitant decrease in antibiotic resistance.

Adaptive Gene Amplification in the lac System

The suggestion that amplification might be an adaptive response to selective conditions (25, 134) was confirmed by the demonstration of adaptive gene amplification in the *lac* system (48). Gene amplification had previously been offered as a possible explanation for adaptive mutation (3): it was proposed that amplification of the mutant *lac* locus to a sufficiently high copy-number during selective growth would increase the likelihood of true reversion events without the need for an increase in mutability. However, Hastings *et al.* (48) showed that amplification at *lac* was induced, and not merely selected, in response to the selective pressure, and occurred in parallel with adaptive point mutation.

ANTIBIOTIC RESISTANCE IN BACTERIA

General Mechanisms

The genetic changes associated with the evolution of antibiotic resistance are well documented (21) and can be classified according to whether they involve lateral gene transfer or chromosomal mutagenesis. However, some pathogenic bacteria may be incapable of exogenous DNA uptake (121) or may occupy specific environments within the host that restrict the opportunities for the acquisition of novel, transmissible genetic elements. For example, the sterility of the ecological niche occupied by M. *tuberculosis*, coupled with the apparent absence of natural competence in mycobacteria (23), implies increased roles for rearrangement and point mutation, rather than lateral gene transfer, in the evolution of resistance within this organism – factors which probably apply to most pathogens colonising locations other than such hotspots for genetic exchange as the nasopharynx or gastrointestinal tract. Since the focus of this chapter is on the mechanisms and implications of stationary-phase mutagenesis, the discussion is limited to the generation and consequences of chromosomal mutations conferring or enhancing antibiotic resistance.

Chromosomal Genes Conferring Antibiotic Resistance

Resistance genes are classically associated with the protection of the target from the drug either through the detoxifying activities of antibiotic-modifying enzymes or extrusion of the antibacterial compounds by efflux pumps. However, in addition to these "target-protection" elements, at least two other classes of genes are associated with the emergence of antibiotic-resistant mutants (70): target-structural genes that are involved in the synthesis and positioning of the antibiotic target, and target-access genes, which are involved in the access of the antibiotic to the cellular target and include enzymes required for the activation of pro-drugs. Mutations in any of the different genes might result in a resistant phenotype, and the likelihood that an antibiotic will invoke several pathways in the bacterial cell means a variety of genes could be involved in resistance to a single antibiotic.

MUTATION TO ANTIBIOTIC RESISTANCE

Antibiotic–Bacterium Interactions

Depending on the specific antibiotic–bacterium interaction at a given antibiotic concentration, resistance can result from single gene mutations (independent mutations) or may require mutations in several genes (cooperative mutations) (70). While single mutations in any of the target-structural, -access, or -protection genes might result in low-level antibiotic resistance, high-level resistance often requires mutations in more than one gene; independent mutations, therefore, tend to be selected at low antibiotic concentrations, whereas cooperative mutations are favoured at higher concentrations. The level of antibiotic resistance may also be variably affected by mutations in different loci, and the stable maintenance of heterogeneous antibiotic resistance expression classes within bacterial populations has been previously demonstrated (30). Often, the observed phenotype will not necessarily reflect the same genotype in all selected mutants, so that a calculated "phenotypic" mutation rate might result from several different "genotypic" mutation events. Furthermore, the mutation rate for a given antibiotic may change depending on its concentration during selection, and it is possible that different rates and resistant mutants may be obtained at discontinuous intervals along the spectrum of applied concentrations (60).

Fitness Cost of Resistance

Antibiotic resistance is often associated with a fitness cost, and the ability of bacteria to compensate for this cost genetically determines the frequency of resistant mutants within a population. Bacterial populations constituting clinically latent infections, by entering into prolonged stationary phase or being confined in isolated lesions or body cavities, may be shielded from the selective pressure of the antibiotic originally causing the mutation. In these circumstances, the fitness cost might be avoided by a reversion to the wild type, antibiotic-sensitive genotype. However, studies have shown that evolution in the absence of the selective antibiotic commonly results in the acquisition of mutations that ameliorate the costs of resistance, rather than higher-fitness, drug-sensitive revertants (7, 118). Furthermore, the nature of the mutations to reduce fitness costs may be environment-dependent, with compensatory mutations apparently arising by different mechanisms within and outside the host (8).

RATES OF MUTATION TO ANTIBIOTIC RESISTANCE

Dependence on the Environment

The environment in which bacteria grow determines the level of variability within a bacterial population (70). In structured and compartmentalised environments, such as might apply in clinically latent infections, the opportunity for direct competition between bacteria is minimised, which allows all alleles capable of surviving the selective pressure to be stably maintained within the population. Growth on structured habitats has been shown to increase the variability of bacterial populations and enable an accelerated evolutionary response to applied environmental pressures (63). The heterogeneity of the environments frequently colonised by bacteria during an infection, whether attached to surfaces or within host cells, suggests that varying degrees of resistant phenotypes might emerge. The broad range of antibiotic susceptibility profiles in *P. aeruginosa* strains isolated from the sputa of cystic fibrosis patients (78) confirms the notion that a wider variety of antibiotic resistant mutants are selected during infection than can be estimated from *in vitro* assays.

The size of the selective habitat and the bacterial population size are factors that can also impact the mutation rate (70). Hypermutable alleles are more likely to be fixed in large populations (131), and a subpopulation of mutators may be beneficially maintained in bacterial populations challenged by the changing, stressful environments within a host. Intuitively, it seems likely that the accelerated evolution conferred on pathogenic bacterial clones by both a higher mutation rate and selection in densely populated, structured habitats may allow bacteria to overcome the challenge posed by antibacterial drugs.

Stresses Encountered *in vivo*

As already noted, physiological conditions such as the availability of a given carbon source (50) or general bacterial stress (32) may regulate the mutation rate in bacteria. The recurrent stresses encountered by pathogenic bacteria attempting to colonise novel environments *in vivo* have led to the suggestion that mutation frequencies in the course of an infective process are probably much higher than those determined by *in vitro* analyses (70). This would apply particularly to pathogens involved in chronic infections, where bacteria are subject to prolonged stress. Unfortunately, the assumption that a lethal selector, such as a bactericidal antibiotic, will eliminate a bacterial population before it has been able to enter prolonged stationary phase has meant that assays of adaptive mutagenesis have tended to utilise a non-lethal selective medium. As a result, regulation of the emergence of antibiotic-resistant mutants under conditions of stress has been inadequately investigated.

Antibiotic Concentration Gradients

The evolution of low-level resistant variants (such as the extended-spectrum TEM β-lactamases) implies the occupation of particular niches within the host in which the locally available concentration of an antibiotic (subject to a chemical gradient) selects for the different variants (5). In many cases, the specific antibiotic concentrations inhibiting the variant and wild-type strains may be extremely close, resulting in a narrow selective concentration range (5). The narrowness of the selective range often results in some low-level resistant mutants being considered clinically irrelevant. However, the ability of a given antibiotic concentration (even if very low) to select for a particular resistance mutation (even if almost neutral) has profound implications for the generation of diverse microbial populations (4). The evolution of high-level resistant mutants may be enhanced *in vivo* in populations in which low-level resistant mutants have been selected and maintained in various compartments; in fact, the role of low-level resistant variants may be analogous to that of multidrug efflux pump systems in the emergence of antibiotic-resistant mutants. Expression of the efflux system genes is usually downregulated and even when constitutively expressed produces only minor variations in the MIC values of antibiotics. However, the idea that mutations in efflux pump systems, in combination with mutations in other antibiotic targets, are the basis of clinically relevant antibiotic-resistant phenotypes has recently gained credence, as demonstrated in the acquisition of quinolone resistance (69).

Stress-Induced Mutagenesis in the Evolution of Antibiotic Resistance?

A direct consequence of the presence of antibiotic concentration gradients *in vivo* is the conclusion that lethal antibiotics might cause non-lethal stress. The importance of stress-related mutagenesis in the emergence of resistant bacteria is becoming increasingly evident, particularly when the selective pressure applied is bacteriostatic. The rate of resistance to quinolones is increased through their use in *E. coli* (109), apparently through induction of the SOS mutagenic response (100), while exposure to streptomycin results in a hypermutable phenotype (106). MDR mutants emerge at an enhanced frequency in *P. aeruginosa* under prolonged challenge with tetracycline (1). Similarly, the forward mutation rate to rifampin resistance increases in *Salmonella enterica* serotype Typhimurium under starvation conditions (50). Mutation to pyrazinamide resistance in *M. tuberculosis* might also be stress-related

(9, 81), given the hostility of the environment in which this drug is believed to mediate its anti-tubercular effects (80). In general, the detection of resistant phenotypes after prolonged exposure to the specific antibiotic is reminiscent of adaptive mutagenesis as defined in the other model systems and hints at the possible involvement of inducible mutator polymerases in such cases.

Stress-Induced Mutagenesis in Genotypic Switching Mechanisms?

Phase variation is a strategy employed by pathogens such as *Bordetella* (142), *Campylobacter* (96), *Haemophilus* (141), *Mycoplasma* (133), and *Neisseria* (107) to rapidly alter their antigenic characteristics and so evade host defences. As noted above, strand-slippage mechanisms allow these organisms to switch between high and low levels of expression of particular groups of genes, the elevated mutation rates being tightly regulated and specific for the hypervariable regions. Analogous switching mechanisms have also been implicated in the emergence of antibiotic-resistant populations (70). In fact, switching between phenotypes of chloramphenicol resistance and susceptibility has been described for isolates of *P. mirabilis* (19) and *Agrobacterium radiobacter* (68), resulting in both cases from altered expression of a chloramphenicol acetyltransferase gene silent in susceptible populations. As in the cases cited above, it is tempting to speculate that Pol IV and its homologues may also play a role in regulating these mutational processes (74).

CONCLUDING COMMENTS

Successful adaptation of bacteria to fluctuating environmental conditions requires the ability to generate diversity. Heritable and local mutation mechanisms, such as those under SOS control, may allow stationary-phase bacteria, especially those comprising clinically latent populations, to adapt to the stresses of nutrient limitation, the host immune response, and antibiotic drugs. The proposal that transient hypermutation is induced in response to environmental conditions does not violate established evolutionary principles. In fact, the processes of mutation and recombination functioning in adaptation are under genetic control, and so are themselves subject to selection. The risks of increased deleterious mutation limit the incidence of constitutive mutators in wild-type bacterial populations, implying an increased role of adaptive mutation strategies, especially those mediated by Pol IV and its homologues, in circumstances in which mutations are beneficial.

Stationary-phase mutagenesis, particularly adaptive mutation mediated by local and environment-dependent mutators, may be implicated in the ascent of antibiotic resistance in clinical bacterial populations. However, resistance generally impacts negatively on bacterial fitness, and the mutations generated to compensate for this cost may entrench antibiotic resistant phenotypes within a population, even after the selective drug is removed. The implication of "irreversible evolution" is that latent clinical infections, maintained by bacteria in isolated niches in the host, may retain antibiotic resistance for the duration of their existence. Finally, the emergence of antibiotic-resistant mutants within a bacterial population is dependent on the physiology, the genetic complement, and the historical behaviour of bacterial populations, as well as the physical structure of the selective medium, and the in-host antibiotic–bacterium dynamics. Because so many factors are involved, the mutation rate and phenotypic variation generated *in vivo* at the site of infection probably differ greatly from that which can be estimated from *in vitro* simulations in the laboratory. Elucidation of the mutagenesis mechanisms occurring *in vivo*, particularly in terms of the role of *dinB*-like genes therein has therefore emerged as an important area for future investigation.

ACKNOWLEDGEMENTS

VM is supported by an International Research Scholars Grant from the Howard Hughes Medical Institute. We are most grateful to Stephanie Dawes for critically reviewing the manuscript and to Steve Durbach and Helena Boshoff for advice, assistance, and helpful discussions.

NOTE ADDED IN PROOF

An amplification-mutagenesis model in which amplification plays an essential role in Lac+ reversion in the *E coli* FC40 system was recently published (Hendrickson, H., Slechta, E. S., Bergthorsson, U., Andersson, D. I., and Roth, J. R. 2002. Proc. Natl. Acad. Sci. USA **99**:2165–2169).

REFERENCES

1. Alonso, A., Campanario, E., and Martinez, J. L. 1999. Emergence of multidrug-resistant mutants is increased under antibiotic selective pressure in *Pseudomonas aeruginosa*. Microbiology **145**:2857–2862.
2. Anderson, R. P., and Roth, J. R. 1977. Tandem genetic duplications in phage and bacteria. Annu. Rev. Microbiol. **31**:473–505.

3. Andersson, D. I., Slechta, E. S., and Roth, J. R. 1998. Evidence that gene amplification underlies adaptive mutability of the bacterial *lac* operon. Science **282**:1133–1135.

4. Baquero, F., and Negri, M.-C. 1997. Selective compartments for resistant microorganisms in antibiotic gradients. Bioessays **19**:731–736.

5. Baquero, F., Negri, M.-C., Morosini, M.-I., and Blazquez, J. 1998. Antibiotic-selective environments. Genetics **27**:S5–S11.

6. Benov, L., and Fridovich, I. 1996. The rate of adaptive mutagenesis in *Escherichia coli* is enhanced by oxygen (superoxide). Mutat. Res. **357**:231–236.

7. Björkman, J., Hughes, D., and Andersson, D. I. 1998. Virulence of antibiotic-resistant *Salmonella typhimurium*. Proc. Natl. Acad. Sci. USA **95**:3949–3953.

8. Björkman, J., Nagaev, I., Berg, O. G., Hughes, D., and Andersson, D. I. 2000. Effects of environment on compensatory mutations to ameliorate costs of antibiotic resistance. Science **287**:1479–1482.

9. Boshoff, H. I. M., and Mizrahi, V. 2000. Expression of *Mycobacterium smegmatis* pyrazinamidase in *Mycobacterium tuberculosis* confers hypersensitivity to pyrazinamide and related amides. J. Bacteriol. **182**:5479–5485.

10. Brègeon, D., Matic, I., Radman, M., and Taddei, F. 1999. Mismatch repair: Genetic defects and down regulation. J. Genet. **78**:21–28.

11. Bridges, B. A. 1995. mutY 'directs' mutation? Nature **375**:741.

12. Bridges, B. A. 1996. Elevated mutation rate in mutT bacteria during starvation: Evidence for DNA turnover? J. Bacteriol. **178**:2709–2711.

13. Bridges, B. A. 1998. The role of DNA damage in stationary phase ('adaptive') mutation. Mutat. Res. **408**:1–9.

14. Bridges, B. A., Sekiguchi, M., and Tajiri, T. 1996. Effect of mutY and mutM/fpg-1 mutations on starvation-associated mutation in *Escherichia coli*: Implications for the role of 7,8-dihydro-8-oxoguanine. Mol. Gen. Genet. **251**:352–357.

15. Brotcorne-Lannoye, A., and Maenhaut-Michel, G. 1986. Role of RecA protein in untargeted UV mutagenesis of bacteriophage lambda: Evidence for the requirement for the *dinB* gene. Proc. Natl. Acad. Sci. USA **83**:3904–3908.

16. Bull, H. J., McKenzie, G. J., Hastings, P. J., and Rosenberg, S. M. 2000. The contribution of transiently hypermutable cells to mutation in stationary phase. Genetics **156**:925–926.

17. Bull, H. J., Lombardo, M. J., and Rosenberg, S. M. 2001. Stationary-phase mutation in the bacterial chromosome: Recombination protein and DNA polymerase IV dependence. Proc. Natl. Acad. Sci. USA **98**:8334–8341.

18. Cairns, J., Overbaugh, J., and Miller, S. 1988. The origin of mutants. Nature **335**:142–145.

19. Charles, I. G., Harford, S., Brookfield, J. F., and Shaw, W. V. 1985. Resistance to chloramphenicol in *Proteus mirabilis* by expression of a chromosomal gene for chloramphenicol acetyltransferase. J. Bacteriol. **164**:114–122.

20. Courcelle, J., Khodursky, A., Peter, B., Brown, P. O., and Hanawalt, P. C. 2001. Comparative gene expression profiles following UV exposure in wild-type and SOS-deficient *Escherichia coli*. Genetics **158**:41–64.

21. Davies, J. E. 1997. Origins, acquisition and dissemination of antibiotic resistance determinants. Ciba Found. Symp. **207**:15–27.

22. Delbrück, M. 1946. Cold Spring Harbor Symp. Quant. Biol. **11**:154.

23. Derbyshire, K. M., and Bardarov, S. 2000. DNA transfer in mycobacteria: Conjugation and transduction. In G. F. Hatfull and W. R. Jacobs (Eds.), *Molecular Genetics of Mycobacteria*, pp. 93–107. ASM Press, Washington, D.C.

24. Dorocicz, I. R., Williams, P. M., and Redfield, R. J. 1993. The *Haemophilus influenzae* adenylate cyclase gene: Cloning, sequence, and essential role in competence. J. Bacteriol. **175**:7142–7149.

25. Echols, H. 1981. SOS functions, cancer and inducible evolution. Cell **25**:1–2.

26. Edlund, T., and Normark, S. 1981. Recombination between short DNA homologies causes tandem duplication. Nature **292**:269–271.

27. Errington, J. 1993. *Bacillus subtilis* sporulation: Regulation of gene expression and control of morphogenesis. Microbiol. Rev. **57**:1–33.

28. Feng, G., Tsui, H.-C. T., and Winkler, M. E. 1996. Depletion of the cellular amounts of the MutS and MutH methyl-directed mismatch repair proteins in stationary-phase *Escherichia coli* K-12 cells. J. Bacteriol. **178**:2388–2396.

29. Field, D., Magnasco, M. O., Moxon, E. R., Metzgar, D., Tanaka, M. M., Wills, C., and Thaler, D. S. 1999. Contingency loci, mutator alleles, and their interactions. Synergistic strategies for microbial evolution and adaptation in pathogenesis. Ann. N.Y. Acad. Sci. **870**:378–382.

30. Figueiredo, A. M., Ha, E., Kreiswirth, B. N., de Lencastre, H., Noel, G. J., Senterfit, L., and Tomasz, A. 1991. *In vivo* stability of heterogeneous expression classes in clinical isolates of methicillin-resistant staphylococci. J. Infect. Dis. **164**:883–887.

31. Finkel, S. E., and Kolter, R. 2001. DNA as a nutrient: Novel role for bacterial competence gene homologs. J. Bacteriol. **183**:6288–6293.

32. Foster, P. L. 1993. Adaptive mutation: The uses of adversity. Annu. Rev. Microbiol. **47**:467–504.

33. Foster, P. L. 1997. Nonadaptive mutations occur in the F′ episome during adaptive mutation conditions in *Escherichia coli*. J. Bacteriol. **179**:1550–1554.

34. Foster, P. L. 1999. Are adaptive mutations due to a decline in mismatch repair? The evidence is lacking. Mutat. Res. **436**:179–184.
35. Foster, P. L. 1999. Mechanisms of stationary phase mutation: A decade of adaptive mutation. Annu. Rev. Genet. **33**:57–88.
36. Foster, P. L., and Trimarchi, J. M. 1994. Adaptive reversion of a frameshift mutation in *Escherichia coli* by simple base deletions in homopolymeric runs. Science **265**:407–409.
37. Friedberg, E. C., Feaver, W. J., and Gerlach, V. L. 2000. The many faces of DNA polymerases: Strategies for mutagenesis and for mutational avoidance. Proc. Natl. Acad. Sci. USA **97**:5681–5683.
38. Gerlach, V. L., Aravind, L., Gotway, G., Schultz, R. A., Koonin, E. V., and Friedberg, E. C. 1999. Human and mouse homologs of *Escherichia coli* DinB (DNA polymerase IV), members of the UmuC/DinB superfamily. Proc. Natl. Acad. Sci. USA **96**:11,922–11,927.
39. Giraud, A., Matic, I., Tenaillon, O., Clara, A., Radman, M., Fons, M., and Taddei, F. 2001. Costs and benefits of high mutation rates: Adaptive evolution of bacteria in the mouse gut. Science **291**:2606–2608.
40. Giraud, A., Radman, M., Matic, I., and Taddei, F. 2001. The rise and fall of mutator bacteria. Curr. Opin. Microbiol. 4:582–585.
41. Grossman, A. D. 1995. Genetic networks controlling the initiation of sporulation and the development of genetic competence in *Bacillus subtilis*. Annu. Rev. Genet. **29**:477–508.
42. Hall, B. G. 1990. Spontaneous point mutations that occur more often when advantageous than when neutral. Genetics **126**:5–16.
43. Hall, B. G. 1998. Adaptive mutagenesis: A process that generates almost exclusively beneficial mutations. Genetica **102/103**:109–125.
44. Haraguchi, Y. and Sasaki, A. 1996. Host-parasite arms race in mutation modifications: Indefinite escalation despite a heavy load? J. Theor. Biol. **183**:121–137.
45. Harris, R. S., Longerich, S., and Rosenberg, S. M. 1994. Recombination in adaptive mutation. Science **264**:258–260.
46. Harris, R. S., Feng, G., Ross, K. J., Sidhu, R., Thulin, C., Longerich, S., Szigety, S. K., Winkler, M. E., and Rosenberg, S. M. 1997. Mismatch repair protein MutL becomes limiting during stationary-phase mutation. Genes Dev. 11:2426–2437.
47. Harris, R. S., Feng, G., Ross, K. J., Sidhu, R., Thulin, C., Longerich, S., Szigety, S. K., Hastings, P. J., Winkler, M. E., and Rosenberg, S. M. 1999. Mismatch repair is diminished during stationary-phase mutation. Mutat. Res. **437**:51–60.

48. Hastings, P. J., Bull, H. J., Klump, J. R., and Rosenberg, S. M. 2000. Adaptive amplification: An inducible chromosome instability mechanism. Cell 103:723–731.

49. Horst, J. P., Wu, T. H., and Marinus, M. G. 1999. *Escherichia coli* mutator genes. Trends Microbiol. 7:29–36.

50. Hughes, D., and Andersson, D. I. 1997. Carbon starvation of *Salmonella typhimurium* does not cause a general increase of mutation rates. J. Bacteriol. 179:6688–6691.

51. Huisman, G. W., Siegele, D. A., Zambrano, M. M., and Kolter, R. 1996. Morphological and physiological changes during stationary phase. In F. C. Neidhardt, R. Curtiss III, J. L. Ingraham, E. C. C. Lin, K. Low Jr., B. Magasanik, W. S. Reznikoff, M. Riley, M. Schaechter, and H. E. Umbarger (Eds.), Escherichia coli *and* Salmonella typhimurium: *Cellular and Molecular Biology*, pp. 1672–1682. ASM Press, Washington, D.C.

52. Ishihama, A. 1997. Adaptation of gene expression in stationary phase bacteria. Curr. Opin. Genet. Dev. 7:582–588.

53. Janbon, G., Sherman, F., and Rustchenko, E. 1999. Appearance and properties of L-sorbose-utilising mutants of *Candida albicans* obtained on a selective plate. Genetics 153:653–664.

54. Kaprelyants, A. S., and Kell, D. B. 1993. Dormancy in stationary-phase cultures of *Micrococcus luteus*: Flow cytometric analysis of starvation and resuscitation. Appl. Environ. Microbiol. 59:3187–3196.

55. Karunakaran, P., and Davies, J. 2000. Genetic antagonism and hypermutability in *Mycobacterium smegmatis*. J. Bacteriol. 182:3331–3335.

56. Kell, D. B., Kaprelyants, A. S., and Grafen, A. 1995. Pheromones, social behaviour and the functions of secondary metabolism in bacteria. Trends. Ecol. Evol. 10:126–129.

57. Kim, S. K., Kaiser, D., and Kuspa, A. 1992. Control of cell density and pattern by intercellular signalling in *Myxococcus* development. Annu. Rev. Microbiol. 46:117–139.

58. Kim, S. R., Maenhaut-Michel, G., Yamada, M., Yamamoto, Y., Masui, K., Sofuni, T., Nohmi, T., and Ohmori, T. 1997. Multiple pathways for SOS-induced mutagenesis in *Escherichia coli*: An overexpression of *dinB/dinP* results in strongly enhancing mutagenesis in the absence of any exogenous treatment to damage DNA. Proc. Natl. Acad. Sci. USA 94:13,792–13,797.

59. Kjelleberg, S., Ed. 1993. *Starvation in Bacteria*. Plenum Press, New York.

60. Kohler, T., Michea-Hamzehpour, M., Plesiat, P., Kahr, A. L., and Pechere, J. C. 1997. Differential selection of multidrug efflux systems by quinolones in *Pseudomonas aeruginosa*. Antimicrob. Agents Chemother. 41:2540–2543.

61. Kolter, R., Siegele, D. A., and Tormo, A. 1993. The stationary phase of the bacterial life cycle. Annu. Rev. Microbiol. **47**:855–874.

62. Kondratyeva, T. F., Muntyan, L. N., and Karavaiko, G. I. 1995. Zinc- and arsenic-resistant strains of *Thiobacillus ferrooxidans* have increased copy numbers of chromosomal resistance genes. Microbiology **141**:1157–1162.

63. Korona, R., Nakatsu, C. H., Forney, L. J., and Lenski, R. E. 1994. Evidence for multiple adaptive peaks from populations of bacteria evolving in a structured habitat. Proc. Natl. Acad. Sci. USA **91**:9037–9041.

64. Kunkel, T. A., and Bebenek, K. 2000. DNA replication fidelity. Annu. Rev. Biochem. **69**:497–529.

65. Lange, R., and Hengge-Aronis, R. 1991. Identification of a central regulator of stationary-phase gene expression in *Escherichia coli*. Mol. Microbiol. **5**:49–59.

66. Lazazzera, B. A. 2000. Quorum sensing and starvation: Signals for entry into stationary phase. Curr. Opin. Microbiol. **3**:177–182.

67. LeClerc, J. E., Li, B., Payne, W. L., and Cebula, T. A. 1996. High mutation frequencies among *Escherichia coli* and Salmonella pathogens. Science **274**:1208–1211.

68. Martinez, J. L., Martinez-Suarez, J., Culebras, E., Perez-Diaz, J. C., and Baquero, F. 1989. Antibiotic inactivating enzymes from a clinical isolate of *Agrobacterium radiobacter*. J. Antimicrob. Chemother. **23**:283–284.

69. Martinez, J. L., Alonso, A., Gómez-Gómez, J. M., and Baquero, F. 1998. Quinolone resistance by mutations in chromosomal gyrase genes. Just the tip of the iceberg? J. Antimicrob. Chemother. **42**:683–688.

70. Martinez, J. L., and Baquero, F. 2000. Mutation frequencies and antibiotic resistance. Antimicrob. Agents Chemother. **44**:1771–1777.

71. Matic, I., Radman, M., Taddei, F., Picard, B., Doit, C., Bingen, E., Denamur, E., and Elion, J. 1997. Highly variable mutation rates in commensal and pathogenic *E. coli*. Science **277**:1833–1834.

72. Matin, A., Auger, E. A., Blum, P. H., and Schultz, J. E. 1989. Genetic basis of stationary-phase survival/recovery in non-differentiating bacteria. Annu. Rev. Microbiol. **43**:293–316.

73. Matthews, P. R., and Stewart, P. R. 1988. Amplification of a section of chromosomal DNA in methicillin-resistant *Staphylococcus aureus* following growth in high concentrations of methicillin. J. Gen. Microbiol. **134**:1455–1464.

74. McKenzie, G. J., and Rosenberg, S. M. 2001. Adaptive mutations, mutator DNA polymerases and genetic change strategies of pathogens. Curr. Opin. Microbiol. **4**:586–594.

75. McKenzie, G. J., Lombardo, M.-J., and Rosenberg, S. M. 1998. Recombination-dependent mutation in *Escherichia coli* occurs in stationary phase. Genetics **149**:1163–1165.

76. McKenzie, G. J., Harris, R. S., Lee, P. L., and Rosenberg, S. M. 2000. The SOS response regulates adaptive mutation. Proc. Natl. Acad. Sci. USA **97**:6646–6651.

77. McKenzie, G. J., Lee, P. L., Lombardo, M.-J., Hastings, P. J., and Rosenberg, S. M. 2001. SOS mutator DNA polymerase IV functions in adaptive mutation and not adaptive amplification. Mol. Cell **7**:571–579.

78. Mereghetti, L., Marquet-van der Mee, N., Loulergue, J., Rolland, J. C., and Audurier, A. 1998. *Pseudomonas aeruginosa* from cystic fibrosis patients: Study using whole cell RAPD and antibiotic susceptibility. Pathol. Biol. **46**:319–324.

79. Metzgar, D., and Wills, C. 2000. Evidence for the adaptive evolution of mutation rates. Cell **101**:581–584.

80. Mitchison, D. A. 1998. How drug resistance emerges as a result of poor compliance during short course chemotherapy for tuberculosis. Int. J. Tuberc. Lung Dis. **2**:10–15.

81. Mizrahi, V., and Andersen, S. J. 1998. DNA repair in *Mycobacterium tuberculosis*. What have we learnt from the genome sequence? Mol. Microbiol. **29**:1331–1339.

82. Modi, R. I., Castilla, L. H., Puskas-Rozsa, S., Helling, R. B., and Adams, J. 1992. Genetic changes accompanying increased fitness in evolving populations of *Escherichia coli*. Genetics **130**:241–249.

83. Modrich, P., and Lahue, R. 1996. Mismatch repair in replication fidelity, genetic recombination, and cancer. Annu. Rev. Biochem. **65**:101–133.

84. Mokkapati, S. K., Fernandez de Henestrosa, A. R., and Bhagwat, A. S. 2001. *Escherichia coli* DNA glycosylase Mug: A growth-regulated enzyme required for mutation avoidance in stationary-phase cells. Mol. Microbiol. **41**:1101–1111.

85. Morita, R. Y. 1993. Bioavailability of energy and the starvation state. In S. Kjelleberg (Ed.), *Starvation in Bacteria*, pp. 1–23. Plenum Press, New York.

86. Moxon, E. R., Rainey, P. B., Nowak, M. A., and Lenski, R. E. 1994. Adaptive evolution of highly mutable loci in pathogenic bacteria. Curr. Biol. **4**:24–33.

87. Moxon, E. R., and Wills, C. 1999. DNA microsatellites: Agents of evolution? Sci. Am. **280**:72–77.

88. Mukamolova, G. V., Kapreylants, A. S., Young, D. I., Young, M., and Kell, D. B. 1998. A bacterial cytokine. Proc. Natl. Acad. Sci. USA **95**:8916–8921.

89. Naas, T., Blot, M., Fitch, W. M., and Arber, W. 1994. Insertion sequence-related genetic variation in resting *Escherichia coli* K-12. Genetics **136**:721–730.

90. Nichols, B. P., and Guay, G. G. 1989. Gene amplification contributes to sulfonamide resistance in *Escherichia coli*. Antimicrob. Agents Chemother. **12**:2042–2048.

91. Notley-McRobb, L., Pinto, R., Seeto, S., and Ferenci, T. 2002. Regulation of *mutY* and nature of mutator mutations in *Escherichia coli* populations under nutrient limitation. J. Bacteriol. **184**:739–745.

92. O'Neal, C. R., Gabriel, W. M., Turk, A. K., Libby, S. J., Fang, F. C., and Spector, M. P. 1994. RpoS is necessary for both the positive and negative regulation of starvation survival genes during phosphate, carbon and nitrogen starvation in *Salmonella typhimurium*. J. Bacteriol. **176**:4610–4616.

93. Ohashi, E., Bebenek, K., Matsuda, T., Feaver, W. J., Gerlach, V. L., Friedberg, E. C., Ohmori, H., and Kunkel, T. A. 2000. Fidelity and processivity of DNA synthesis by DNA polymerase kappa, the product of the human DINB1 gene. J. Biol. Chem. **275**:39,678–39,684.

94. Oliver, A. R., Canton, P., Campo, F., Baquero, F., and Blazquez, J. 2000. High frequency of hypermutable *Pseudomonas aeruginosa* in cystic fibrosis lung infection. Science **288**:1251–1253.

95. Ono, S. 1970. *Evolution by Gene Duplication*. Springer-Verlag, Berlin.

96. Park, S. F., Purdy, D., and Leach, S. 2000. Localized reversible frameshift mutation in the *flhA* gene confers phase variability to flagellin gene expression in *Campylobacter coli*. J. Bacteriol. **182**:207–210.

97. Parrish, N. M., Dick, J. D., and Bishai, W. R. 1998. Mechanisms of latency in *Mycobacterium tuberculosis*. Trends Microbiol. **6**:107–112.

98. Peterson, B. C., and Rownd, R. H. 1985. Drug resistance gene amplification of plasmid NR1 derivatives with various amounts of resistant determinant DNA. J. Bacteriol. **161**:1042–1048.

99. Pham, P., Bertram, J. G., O'Donnell, M., Woodgate, R., and Goodman, M. F. 2001. A model for SOS-lesion-targeted mutations in *Escherichia coli*. Nature **409**:366–370.

100. Phillips, I., Culebras, E., Moreno, F., and Baquero, F. 1987. Induction of the SOS response by new 4-quinolones. J. Antimicrob. Chemother. **20**:631–638.

101. Radman, M. 1975. SOS repair hypothesis: Phenomenology of an inducible DNA repair which is accompanied by mutagenesis. Basic Life Sci. **5A**:355–367.

102. Radman, M. 1999. Mutation: Enzymes of evolutionary change. Nature **401**:866–869.

103. Radman, M., Taddei, F., and Matic, I. 2000. Evolution-driving genes. Res. Microbiol. **151**:91–95.

104. Rebeck, G. W., and Samson, L. 1991. Increased spontaneous mutation and alkylation sensitivity of *Escherichia coli* strains lacking the ogt O6-methylguanine DNA repair methyltransferase. J. Bacteriol. **173**:2068–2076.

105. Reeve, C. A., Amy, P. S., and Matin, A. 1984. Role of synthesis in the survival of carbon-starved *Escherichia coli*. J. Bacteriol. **160**:1041–1046.

106. Ren, L., Rahman, M. S., and Humayun, M. Z. 1999. *Escherichia coli* cells exposed to streptomycin display a mutator phenotype. J. Bacteriol. **181**:1043–1044.

107. Richardson, A. R., and Stojiljkovic, I. 2001. Mismatch repair and the regulation of phase variation in *Neisseria meningitidis*. Mol. Microbiol. **40**:645–655.

108. Riehle, M. M., Bennett, A. F., and Long, A. D. 2001. Genetic architecture of thermal adaptation in *Escherichia coli*. Proc. Natl. Acad. Sci. USA **98**:525–530.

109. Riesenfeld, C., Everett, M., Piddock, L. J., and Hall, B. G. 1997. Adaptive mutations produce resistance to ciprofloxacin. Antimicrob. Agents Chemother. **41**:2059–2060.

110. Romero, D., and Palacios, R. 1997. Gene amplification and genomic plasticity in prokaryotes. Annu. Rev. Genet. **31**:91–111.

111. Rosche, W. A., and Foster, P. L. 1999. The role of transient hypermutators in adaptive mutation in *Escherichia coli*. Proc. Natl. Acad. Sci. USA **96**:6862–6867.

112. Rosenberg, S. M., Longerich, S., Gee, P., and Harris, R. S. 1994. Adaptive mutation by deletions in small mononucleotide repeats. Science **265**:405–407.

113. Rosenberg, S. M. 1997. Mutation for survival. Curr. Opin. Genet. Dev. **7**:829–834.

114. Rosenberg, S. M. 2001. Evolving responsively: Adaptive mutation. Natl. Rev. Genet. **2**:504–515.

115. Roth, J. R., Benson, N., Galitski, T., Haack, K. R., Lawrence, J. G., and Miesel, L. 1996. Rearrangements of the bacterial chromosome: Formation and applications. In F. C. Neidhardt, R. Curtiss, III, J. L. Ingraham, E. C. C. Lin, K. Low Jr., B. Magasanik, W. S. Reznikoff, M. Riley, M. Schaechter, and H. E. Umbarger (Eds.), Escherichia coli *and* Salmonella typhimurium: Cellular and Molecular Biology, pp. 2256–2276. ASM Press, Washington, D.C.

116. Ryan, F. J. 1955. Spontaneous mutation in non-dividing bacteria. Genetics **40**:726–738.

117. Ryan, F. J. 1959. Bacterial mutation in stationary phase and the question of cell turnover. J. Gen. Microbiol. **21**:530–549.

WARNER AND MIZRAHI

118. Schrag, S., and Perrot, V. 1996. Reducing antibiotic resistance. Nature **381**:120–121.
119. Siegele, D. A., and Kolter, R. 1992. Life after log. J. Bacteriol. **174**:345–348.
120. Smeulders, M. J., Keer, J., Speight, R. A., and Williams, H. D. 1999. Adaptation of *Mycobacterium smegmatis* to stationary phase. J. Bacteriol. **181**:270–283.
121. Solomon, J. M., and Grossman, A. D. 1996. Who's competent and when: Regulation of natural genetic competence in bacteria. Trends Genet. **12**:150–155.
122. Sonti, R. V., and Roth, J. R. 1989. Role of gene duplications in the adaptation of *Salmonella typhimurium* to growth on limiting carbon sources. Genetics **123**:19–28.
123. Stark, G. R., and Wahl, G. M. 1984. Gene amplification. Annu. Rev. Biochem. **53**:447–491.
124. Storchova, Z., and Vondrejs, V. 1999. Starvation-associated mutagenesis in yeast *Saccharomyces cerevisiae* is affected by Ras2/cAMP signalling pathway. Mutat. Res. **431**:59–67.
125. Taddei, F., Matic, I., and Radman, M. 1995. cAMP-dependent SOS induction and mutagenesis in resting bacterial populations. Proc. Natl. Acad. Sci. USA **92**:11,736–11,740.
126. Taddei, F., Halliday, J. A., Matic, I., and Radman, M. 1997. Genetic analysis of mutagenesis in aging *Escherichia coli* colonies. Mol. Gen. Genet. **256**:277–281.
127. Taddei, F., Radman, M., Maynard-Smith, J., Toupance, B., Gouyon, P. H., and Godelle, B. 1997. Role of mutator alleles in adaptive evolution. Nature **387**:700–702.
128. Tang, M., Pham, P., Shen, X., Taylor, J. S., O'Donnell, M., Woodgate, R., and Goodman, M. F. 2000. Roles of *E. coli* DNA polymerases IV and V in lesion-targeted and untargeted SOS mutagenesis. Nature **404**:1014–1018.
129. Tang, M., Shen, X., Frank, E. G., O'Donnell, M., Woodgate, R., and Goodman, M. F. 1999. UmuD'$_2$C is an error-prone DNA polymerase, *Escherichia coli* pol V. Proc. Natl. Acad. Sci. USA **96**:8919–8924.
130. Taverna, P., and Sedgwick, B. 1996. Generation of an endogenous DNA-methylating agent by nitrosation in *Escherichia coli*. J. Bacteriol. **178**:5105–5111.
131. Tenaillon, O., Toupance, B., Le Nagard, H., Taddei, F., and Godelle, B. 1999. Mutators, population size, adaptive landscape and the adaptation of asexual populations of bacteria. Genetics **152**:485–493.
132. Tenaillon, O., Taddei, F., Radman, M., and Matic, I. 2001. Second-order selection in bacterial evolution: Selection acting on mutation and recombination rates in the course of adaptation. Res. Microbiol. **152**:11–16.

133. Theiss, P., and Wise, K. S. 1997. Localized frameshift mutation generates selective, high-frequency phase variation of a surface lipoprotein encoded by a mycoplasma ABC transporter operon. J. Bacteriol. **179**:4013–4022.

134. Tlsty, T. D., Albertini, A. M., and Miller, J. H. 1984. Gene amplification in the lac region of *E. coli*. Cell **37**:217–224.

135. Torkelson, J., Harris, R. S., Lombardo, M.-J., Nagendran, J., Thulin, C., and Rosenberg, S. M. 1997. Genome-wide hypermutation in a subpopulation of stationary-phase cells underlies recombination-dependent adaptive mutation. EMBO J. **16**:3303–3311.

136. Vulić, M., and Kolter, R. 2001. Evolutionary cheating in *Escherichia coli* stationary phase cultures. Genetics **158**:519–526.

137. Wagner, J., Gruz, P., Kim, S. R., Yamada, M., Matsui, K., Fuchs, R. P., and Nohmi, T. 1999. The *dinB* gene encodes a novel *E. coli* DNA polymerase, DNA pol IV, involved in mutagenesis. Mol. Cell **4**:281–286.

138. Wagner, J., and Nohmi, T. 2000. *Escherichia coli* DNA polymerase IV mutator activity: Genetic requirements and mutational specificity. J. Bacteriol. **182**:4587–4595.

139. Walker, G. C., Smith, B. T., and Sutton, M. D. 2000. The SOS response to DNA damage. In G. Storz and R. Hengge-Aronis (Eds.), *Bacterial Stress Responses*, pp. 131–144. ASM Press, Washington, D.C.

140. Watson, S. P., Clements, M. O., and Foster, S. J. 1998. Characterisation of the starvation-survival response of *Staphylococcus aureus*. J. Bacteriol. **180**:1750–1758.

141. Weiser, J. N., Love, J. M., and Moxon, E. R. 1989. The molecular mechanism of phase variation of *H. influenzae* lipopolysaccharide. Cell **59**:657–665.

142. Willems, R., Paul, A., van der Helde, H. G., ter Avest, A. R., and Mooi, F. R. 1990. Fimbrial phase variation in *Bordatella pertussis*: A novel mechanism for transcriptional regulation. EMBO J. **9**:2803–2809.

143. Yagi, Y., and Clewell, D. B. 1980. Amplification of the tetracycline resistance determinant of plasmid pAMa1 in *Streptococcus faecalis*: Dependence on host recombination machinery. J. Bacteriol. **143**:1070–1072.

144. Zambrano, M. M., and Kolter, R. 1993. *Escherichia coli* mutants lacking NADH dehydrogenase-1 have a competitive disadvantage in stationary phase. J. Bacteriol. **175**:5642–5647.

145. Zambrano, M. M., and Kolter, R. 1996. GASPing for life in stationary phase. Cell **86**:181–184.

146. Zambrano, M. M., Siegele, D. A., Almirón, M., and Tormo, A. 1993. Microbial competition: *Escherichia coli* mutants that take over stationary phase cultures. Science **259**:1757–1760.

147. Zhang, Y., Yuan, F., Xin, H., Wu, X., Rajpal, D. K., Yang, D., and Wang, Z.

WARNER AND MIZRAHI

2000. Human DNA polymerase kappa synthesizes DNA with extraordinarily low fidelity. Nucleic Acids Res. **28**:4147–4156.

148. Zhao, J., and Winkler, M. E. 2000. Reduction of GC→TA transversion mutation by overexpression of MutS in *Escherichia coli* K-12. J. Bacteriol. **182**: 5025–5028.

149. Zinser, E. R., and Kolter, R. 1999. Mutations enhancing amino acid catabolism confer a growth advantage in stationary phase. J. Bacteriol. **181**: 5800–5807.

CHAPTER 6

Biofilms, dormancy and resistance

Anthony W. Smith and Michael R. W. Brown

INTRODUCTION

Extent of the Problem in Medicine

Biofilm growth almost always leads to a massive increase in resistance to antibiotics and biocides by orders of magnitude compared with cultures grown in suspension (planktonic) in conventional liquid media. Currently, there is no generally agreed upon mechanism to account for this broad resistance to chemical agents. We suggest that dormancy, related to the general stress response and associated survival responses, offers an explanation for the overall general resistance of biofilm microbes.

A bacterial biofilm is typically defined as a population of cells growing as a consortium on a surface and enclosed in a complex exopolymer matrix. Commonly in the wider environment but less so in infections, the population is mixed and also of heterogeneous physiologies (22). This is partly because of nutrient (including oxygen) gradients across the biofilm (92). It seems likely that biomasses not associated with a surface may also exhibit biofilm properties, because of the influence of cell density on microbial physiology (12, 37). The biofilm mode of growth contributes to resistance to host defences and, within the biofilm, at least some of the cells are phenotypically highly resistant to antibiotics and biocides (58, 60). Medical examples are numerous and include dental biofilms and also burns and urinary tract and lower respiratory infections. The carrier state for several infections involves pharyngeal colonisation. Implanted devices are a major source of biofilm infection, e.g., catheters and pacemakers, as are orthopaedic and other prostheses (4). In a recent survey of hospitals in England, two thirds of bacteraemias were associated with an intracellular device or device-related infections, and their management extended hospital stays by 11 days (74).

Surface Adhesion and Biofilm Formation

More detailed accounts of the physics of attachment can be studied elsewhere (4, 41). Also, the role of specific adhesin/receptor binding to host cell surfaces, with consequent complex signal transduction cascades in the host cell, is reviewed elsewhere (8, 50).

For medical purposes, there is no such thing as an unadulterated surface. Any surface, synthetic or otherwise, rapidly becomes coated with constituents of the local environment: first water and salt ions, then organic material such as the proteins albumin or fibronectin. This conditioning film exists before the arrival of the first microbe. Next, there is a weak and reversible contact between microbe and conditioning film resulting from Brownian motion, gravitation, diffusion or microbial motility and involving electrostatic interactions. The surface interaction is a function of the cell surface (determined by the cell physiology) and the nature of the film. For example, recent work with an *S. aureus* mutant bearing a stronger negative charge due to the lack of D-alanine esters in its teichoic acids showed that it could no longer colonise polystyrene or glass surfaces, highlighting the contribution of electrostatic forces to biofilm formation, particularly on medical devices (43). Charge attraction or repulsion could also contribute to interaction between antibiotics and the cell surface. Specific interactions with bacterial surface structures can also be important in establishing a biofilm. For example, flagella and pilus-mediated twitching motility are required/important for *E. coli* and *P. aeruginosa* biofilms (69, 73). The third step is when the adsorption becomes irreversible. This is partly due to surface appendages overcoming the repulsive forces between the two surfaces and also because of sticky exopolymers secreted by the cells. Commonly the entire biofilm may be coated with a hydrophilic exopolymer (the glycocalyx), which is itself a complex and dynamic structure (87). When the host is unable to opsonise this hydrophilic glycocalyx, the entire biofilm is resistant to phagocytosis (71). The fourth, overlapping, stage is when the biofilm population increases, typically as a result of adherent cells replicating, but including a minor contribution from fresh cells adhering to the biofilm (2, 39). In Staphylococcal species, commonly associated with device-related infections, biofilm formation requires cell–cell adhesion mediated by the *ica* locus following adhesion to the substratum (23). Sub-inhibitory concentrations of tetracycline and the semi-synthetic streptogramin antibiotic quinupristin–dalfopristin enhanced *ica* expression and biofilm formation (76).

The biofilm structure/phenotype depends on numerous factors, notably the organism and its physiology, the substratum, the surrounding nutrient environment and the rate of flow of any liquid over its surface (21, 53). The resulting biofilms may vary from sparse amorphous masses to highly structured consortia with mushroom-like cell stacks surrounded by channels with rapid aqueous movement (22). Signal molecules (see below) which influence cell physiology including virulence and the general stress response (and thus dormancy) have also been shown to influence biofilm structure.

This chapter is largely restricted to gross biofilms involved in infection. The initial step of biofilm formation, i.e., adhesion, is relevant to infection in three different ways. There is adhesion to (1) host cell surfaces, (2) extracellular matrix and plasma components and (3) the synthetic surfaces of medical devices (4, 8, 59). The latter two are most likely to lead to gross biofilms with potential to avoid phagocytosis and also for phenotypic antibiotic resistance.

Biofilm Models

There are numerous biofilm culture models reviewed elsewhere (3, 29, 54, 80, 95). Valid comparisons between biofilm and planktonic cultures are difficult to make, especially if key parameters such as growth rate and/or specific nutrient limitation are uncontrolled and vary between biofilm and planktonic culture. Much published work compares biofilm cultures in one physiological state with planktonic cultures in a different state and often in a different medium. A model which controls biofilm growth rate in a defined, nutrient-limited medium consists of surface-growth on the underside of a bacteria-proof cellulose membrane (38). The membrane is perfused with fresh, defined medium from the sterile side and cells eluted from the biofilm are collected. Growth rate of the nutrient-limited, adherent population is controlled by rate of perfusion of fresh medium. A method of controlling density and nutrient limitation of biofilm growth consists of membrane culture on the surface of defined agar medium, growth limited by any specified major nutrient (16, 28). These methods enable comparison of planktonic and biofilm cultures at similar stages of growth and under similar eventual nutrient restriction.

CONVENTIONAL CANDIDATES TO EXPLAIN RESISTANCE TO ANTIBIOTICS AND BIOCIDES: "THE USUAL SUSPECTS"

In a recent review, the often-cited mechanisms of antibiotic resistance have been referred to as "the usual suspects" (58). These are outlined below; however, clarification of the concept of resistance could be helpful. In the case of populations of bacteria involved in, say, a chronic infection or a device-associated one, then antibiotic eradication of the infection requires elimination of all the bacteria, typically assisted by the host defences. Otherwise infection re-occurs. In other words, biofilm resistance can be determined by the susceptibility of the most resistant cell. It is not the case that all cells within a biofilm are always highly resistant (9, 58). But the most resistant members of a biofilm population are virtually always orders of magnitude more resistant than similar members of a planktonic population. In the context of biofilm resistance it is necessary to examine if any of the conventional mechanisms of antibiotic resistance play an enhanced role in biofilms, thus contributing to their extraordinary resistance.

Inability to Penetrate the Biofilm

There are numerous papers investigating possible lack of antibiotic/ biocide penetration as an explanation of biofilm resistance (41, 58, 60, 83, 94). Given, in some cases, biofilms consisting largely of stacks of cells with flowing aqueous channels (even though coated with glycocalyx), impenetrability seems highly unlikely (66), a finding confirmed with biofilms of *Klebsiella pneumoniae* (5). Indeed, the channels between stacks can be sufficiently wide for protozoa to graze between them (Costerton and Davies, personal communication, 2002).

Nevertheless, where the antimicrobial agent either reacts chemically with components of the exopolymer or is significantly adsorbed by these typically anionic polymers, the net effect is as if there is a penetration barrier. There will be a similar effect if such interactions occur with cells, perhaps dead ones, in the outer parts of the biofilm. Heterogeneity has also been given as a reason for biofilm resistance. But, in terms of eradication/sterilisation, resistance is caused by the most resistant members of the biofilm. Hence the question still remains as to the mechanism. Heterogeneity *per se* is not a mechanism. Given that biofilms are indeed typically heterogeneous, these generalisations do not preclude the possibility of areas in a biofilm where diffusion is restricted (58) and is perhaps coupled with the presence of dormant cells.

Enzyme Inactivation of Antimicrobials

Antibiotics and other antimicrobial agents can commonly induce the production of inactivating enzymes in microbes. The relatively large amounts of antibiotic-inactivating enzymes such as β-lactams which accumulate within the glycocalyx produce concentration gradients of antimicrobial across it, and the underlying cells have been shown to be thus protected (7, 42), although in other systems enzyme inactivation appears not to contribute to the resistance of biofilm bacteria (5).

Efflux

The contribution of efflux systems to the resistance of bacteria to antimicrobial agents has been studied extensively in recent years (58). Several classes of antibiotic are substrates for the pumps and include the tetracyclines, macrolides, β-lactams and fluoroquinolones (89). Most studies of the contribution of efflux systems to antibiotic resistance have been performed on planktonic cells, i.e., cells in suspension. However, recent studies by Gilbert and co-workers have addressed the contribution of the multiple antibiotic resistance (MAR) operon and AcrAB in biofilms of *E. coli* (61). The Mar operon is present in a number of Gram-negative bacteria and the antibiotic resistance phenotype is mediated by upregulation of AcrAB (65). Mutants deleted for Mar showed sensitivity to the fluoroquinolone antibiotic ciprofloxacin similar to that of wild-type cells grown in a biofilm, whereas a constitutive mar mutant showed decreased susceptibility. Isolates in which the acrAB efflux pump was deleted also showed similar sensitivity, whilst constitutive expression of acrAB protected biofilms at low antibiotic concentrations but not high, leading the authors to conclude that ciprofloxacin resistance in *E. coli* biofilms is not mediated by upregulation of the mar and acrAB operons (61). In a comprehensive study of multi-drug efflux pumps in *P. aeruginosa*, none of the four systems present in the genome contributed to resistance in a biofilm (26). Temporal and spatial analyses using fusions to *gfp* indicated that expression of *mexAB-oprM* and *mexCD-oprJ* decreased over time in the developing biofilm, with maximal expression occurring at the biofilm substratum.

The contribution of efflux pumps to quorum sensing and biofilm formation is now becoming apparent and thus there is potential for efflux pumps to contribute to resistance of biofilm cells through mechanisms relating to stress responses and dormancy rather than drug efflux *per se*. LasI mutants of *P. aeruginosa* deficient in production of the N-(3-oxododecanoyl) homoserine lactone (3OC12-HSL) autoinducer molecule formed flat, undifferentiated biofilms that were sensitive to treatment with sodium dodecyl sulphate (24). Evidence suggests that this quorum-sensing molecule is a substrate for the MexAB–OprM multi-drug efflux system, since mutants which hyperexpress this system showed reduced levels of extracellular virulence factors known to be regulated by quorum sensing (34). Also, a defined mutant lacking the pump accumulated more 3OC12-HSL, comparable with wild-type cells treated with cytoplasmic membrane proton gradient inhibitors (70). The temporal events integrating biofilm formation, production of virulence factors and the general stress response await further investigation.

Repair

Operation of repair systems possibly enhanced within a biofilm could contribute to the decreased susceptibility of cells within a biofilm to antimicrobial agents. Enzymes involved with detoxification of reactive oxygen species, notably superoxide dismutase and catalase, have been the most extensively studied. Hassett and co-workers have shown that levels of the manganese- and iron-cofactored superoxide dismutases and the major catalase KatA are decreased in mutants of *P. aeruginosa* devoid of one or both quorum-sensing molecules grown planktonically, with a concomitant increase in sensitivity to hydrogen peroxide and phenazine methosulphate (45). Perhaps surprisingly, biofilm-grown cells had less catalase activity and yet were more resistant to hydrogen peroxide than their planktonic counterparts. Catalase levels were even lower in quorum-sensing deficient mutants and yet they were also resistant to hydrogen peroxide. One resistance mechanism for catalase appears to be prevention of hydrogen peroxide penetration fully into the biofilm (84).

The susceptibility to reactive oxygen species may also be related to repair of sub-lethal injury following antimicrobial treatment. It has been proposed that sub-lethal injury by antimicrobial agents leads to an imbalance in anabolism and catabolism and a burst of damaging oxygen free-radical production (30) on attempting to recover treated cultures. Viable competitive microflora at high density can protect exponential phase cells of another organism at lower density from the lethal effects of heat (chemical antimicrobials have not been tested). The authors propose that this addition creates an immediate reduction in the oxygen tension of the culture and oxidative metabolism is reduced (30). Although not tested, variation in oxygen tension gradients within a biofilm, together with the varying metabolic activities within a mixed microbial biofilm, could be conceived to contribute to resistance.

SURVIVAL AND THE RESPONSE TO STRESS: WILL THE REAL CULPRIT PLEASE STAND UP?

In the natural environment, survival of Gram-positive *Bacillus* and *Clostridium* species occurs via resistant spores. For non-sporulating bacteria, especially but not exclusively Gram-negative ones, there is a growing literature on an aspect of the behaviour of the stationary phase of planktonic bacteria known as the general stress response (GSR) (48) (see Chapter 2). This stress response has been implicated directly (37) in chronic infection involving biofilms and could clearly occur in circumstances where high density and quorum sensing have been reported (79, 85, 93). Whilst the emphasis here is on Gram-negative bacteria, the reader is reminded that there are well characterised stress response systems in Gram-positive bacteria. Systems under the control of alternative sigma factors include *Bacillus subtilis* (77), *Staphylococcus aureus* (18, 19), where biofilm formation is also affected (75) and *Mycobacterium tuberculosis* (27).

The General Stress Response (GSR)

The GSR involves a late log and stationary-phase cascade during which structures are protected and the cells become quiescent. It resembles sporulation in its physiological consequences. The result is an ability to survive prolonged periods of nutrient starvation and multiple environmental stresses, such as heat, oxidising agents and hyperosmolarity. Unlike sporulation, the stationary-phase response does not involve an all-or-nothing switch nor an irreversible commitment. Also, some genes involved exhibit expression which is inversely related to growth rate and are already partially induced under conditions of slow growth (47). The final, slow/non-replicating stages have been described as quiescent, resting or dormant. The expression of many proteins is regulated on entry into stationary phase, at which point a core set is induced regardless of the cause of cessation of growth, for example, the nature of the depleted nutrient (62). In *E. coli* the *rpoS*-encoded sigma factor σ^s is a master regulator of the GSR (47). There are numerous papers showing a general tendency for nutrient depletion and slow or no growth to be associated with antibiotic and biocide resistance (10, 14, 15). In retrospect, it seems probable that, in addition to the consequences of adaptation to the specific nutrient depletion and reduced growth rate, the GSR plays a major role in resistance. There is also evidence that specific nutrient depletion has major effects on sensitivity of microorganisms to host defences (6, 36). Nevertheless, there is as yet little work on the effects of the GSR *per se* on susceptibility to host defences (67) or antibiotics (63). Its role in biofilm formation is not clear, although *E. coli* biofilm density was reduced in an *rpoS* mutant grown in a modified Robbins device (1). Whilst Gram-negative bacteria are not challenged with glyocopeptide antibiotics because of the permeability barrier of the outer membrane, D-alanine-D-alanine dipeptidase, part of the five-gene cluster involved in reprogramming peptidoglycan biosynthesis toward D-alanine-D-lactate and decreased susceptibility to vancomycin, is transcribed in stationary phase by RpoS (57). In *E. coli* it is thought that the

D-alanine could be used as an energy source for cell survival under starvation conditions.

Although *rpoS* is a global regulator of the GSR in many gram-negative species, it is not solely responsible for survival/resistance. Work on *rpoS* mutants confirms that other resistance mechanisms exist (52). Also, transcription, translation and proteolysis of *rpoS* is affected by numerous influences including ppGpp and the stringent response (17), polyphosphate (78), OxyR (72), SOS (88) and cAMP–CRP (81). Clearly, responses to stresses involve overlapping networks with high degrees of interaction.

Biofilms and Quorum Sensing

Quorum-sensing (QS) systems comprise a transcriptional activator protein that acts in concert with a low-molecular-weight autoinducer (AI) signalling molecule to increase expression of target genes (see Chapter 4). As the cell population density increases, so does AI density, providing a means to monitor cell density. Whilst early studies were performed on cells growing in planktonic culture, attention is now being focused on the role of QS in biofilms. By definition, cell density will be high in a compact, adherent biofilm population, and consequently, relatively small biofilm populations probably demonstrate signal-driven, stationary-phase survival responses which equivalent numbers of free-growing planktonic counterparts would not (41). This could partially explain the general high resistance of biofilm organisms to exogenous stress.

As described above, *P. aeruginosa lasI* mutants which do not produce the N-(3-oxododecanoyl) homoserine lactone (3OC12-HSL) AI molecule fail to form differentiated biofilms and are sensitive to biocide inactivation (24). *P. aeruginosa* produces two well characterised AI molecules, 3OC12-HSL and *N*-butyryl-L-homoserine lactone (C4-HSL). In conventional planktonic culture *P. aeruginosa* produces 3OC12-HSL at a rate between 3 and 10 times that of C4-HSL, whereas more C4-HSL was produced by biofilm-grown cells (79). Greater C4-HSL levels were noted in extracts of sputum from cystic fibrosis patients, indicating that bacteria are perhaps growing as a biofilm in the lungs of these patients (79).

Clearly, signal diffusion will be influenced by the nature of the biofilm matrix and of the substratum. Thus, an impermeable substratum for biofilm growth would concentrate any signal, while the degree of hydrophobicity of a boundary could influence entrapment or diffusion, depending on the chemistry of the signal. A hydrophobic signal could be trapped as aggregates/ micelles within cell exopolymer and maintain an equilibrium concentration of hydrophobic monomer close to the cell, while hydrophilic molecules could diffuse away. Consistent with this hypothesis, expression of *P. aeruginosa* AI synthase genes was greatest at the interface with the impermeable substratum (25), permitting rapid amplification of cell-density-dependent responses since the AI synthase genes themselves are subject to autoregulation. The N-acyl substituted HSLs vary with respect to the state of oxidation at the C-3 position and the fatty-acid chain length. Relative hydrophobicity/lipophilicity for a bioactive compound can be predicted by the parameter log P. Optimum permeation through a Gram-negative envelope is commonly at about log P of 4. Longer chain length AI molecules, such as 3OC12-HSL from *P. aeruginosa*, having a log P of approximately 3, would be predicted to enter or exit cells by simple diffusion (49) and may well form aggregates under the appropriate conditions. However, in at least the case of *P. aeruginosa*, 3OC12-HSL requires active efflux (70).

Further evidence for the complex interplay of regulatory factors comes from the hierarchical link between RpoS expression and quorum sensing, with recent evidence indicating negative regulation of C4-HSL synthesis (91). How these events might integrate within a biofilm awaits further investigation.

Acylated homoserine lactones do not appear to operate in members of the enterobacteriacae such as *E. coli* and *Salmonella enteritidis*, although autoinducer activity has been reported (86) and QS regulates activity of the locus of enterocyte attachment and effacement and intestinal colonisation (82).

A number of QS or cell-density-dependent systems operate in Gram-positive species, although here the AI molecules are typically small peptides. Examples include regulation of streptococcal competence for genetic transformation (46); cell-density control of gene expression and sporulation in *Bacillus* species (56); the sex pheromone systems regulating conjugative plasmid transfer in Enterococcal species, some of which have associated antibiotic resistance (20); and regulation of *Staphylococcus aureus* pathogenicity (68). In most cases, their direct role in modulating susceptibility to antimicrobials has not been reported. However agents such as chloramphenicol and tetracycline at subinhibitory concentration can perturb signalling cascades through inhibition of AI peptide synthesis. Studies with the *agr* quorum-sensing system in *S. aureus* have shown that QS-deficient mutants were more able to form a biofilm on polystyrene than wild-type strains, leading the authors to question the utility of anti-QS molecules to eradicate biofilm infections (90).

Evidence for the direct contribution of AI or quorum-sensing molecules to the antimicrobical susceptibility of bacteria is lacking. Nevertheless, given their central role in virulence factor production and biofilm formation, they are themselves attractive targets for antimicrobial drug design. However, it is already clear that there are fundamental differences in the contribution of these systems to biofilm formation between species. Such differences are not surprising. It is necessary to bear in mind the plethora both of bacterial phenotypes and of potential surfaces for colonisation.

Potential Role of RpoS in Uptake/Adhesion

Although as yet little studied in mature biofilms, *rpoS* has a role in cell uptake/adhesion. Furthermore, in addition to its role in resistance/survival, *rpoS* activity and the subsequent GSR response, it is associated with pathogenicity and possible surface antigen changes. RpoS regulates both chromosomally encoded genes and plasmid-encoded *spv* genes, important in the virulence of *S. typhimurium* (35). RpoS is necessary for the early stages of *Salmonella* uptake by Peyer's patches of the gut (67). Humphrey and co-workers have shown a correlation with *Salmonella enterica* serotype Enteritidis PT4 between heat and acid tolerance and virulence in chickens (51). There is also a recent report that *L. pneumophila rpoS* is required for growth within amoebae although apparently not required for stationary-phase stress resistance (44). The question arises as to whether other pathogens need the general stress response for optimum host uptake, although there are examples where RpoS is either not involved (64) or disadvantageous (55).

SLOW/NO GROWTH – AN ACCOMPLICE TO DORMANCY?

Decreased culture growth rate is associated with decreased susceptibility to almost all chemical antimicrobial agents, some of which have a requirement for replication (11, 14, 60). Also, any stress response by a growing culture is associated with a reduction in growth rate and even growth cessation. This makes it difficult to distinguish the individual contributions to the start of a resistance cascade of growth rate *per se* or/and an enforced *change* in rate, cell density (and/or quorum sensing) and the nature of a nutrient starvation or other stress. In the case of a biofilm, density is high at an early stage relative to the same number of cells growing in a conventional planktonic culture. It is also difficult to make a valid comparison between biofilm cells and planktonic cells when they have been cultured often in different media and harvested in different physiological states. Using growth rate–controlled cultures of *P. aeruginosa, E. coli* and *S. epidermidis,* in both planktonic and biofilm modes of growth, there was a definite growth rate effect. In both cases sensitivity increased with increased growth rate (31–33). However, increases in growth rate caused bigger changes in sensitivity with planktonic cells, indicating factors operating in addition to growth rate. Using chemically and nutritionally defined cultures (13, 16) which ultimately entered stationary phase because of iron-limitation, susceptibility to ciprofloxacin and to ceftazidime was measured along the exponential phase of batch cultures, planktonic and biofilm (28). In both growth modes there were dramatic changes in resistance throughout the exponential phase and before measurable growth reduction for stationary phase. Increases in resistance to both agents occurred in planktonic culture about 3–4 generations before onset of stationary phase and with biofilms about 10 generations before stationary phase. Cell density may well have played a part, although effects were noted well below the commonly recorded quorum-sensing density.

A consideration of the available literature shows that stress responses are always accompanied by a reduction in growth rate. Even static suspensions used for susceptibility assays are from cultures that have ceased growth due to starvation or the presence of inhibitory substances. Also, handling techniques such as centrifugation and re-suspension can contribute to stress (40). Consequently, a reduction in growth rate is an indication of a stress and a specific slow growth rate may well maintain the cells in the initial stages of a stress response. Thus slow growth rate is an accomplice but not the main culprit.

CONCLUDING COMMENTS

The general resistance of biofilms is clearly phenotypic. The usual suspects – lack of antibiotic penetration, inactivation, efflux and repair – are probably not guilty. The suspicion that reduced growth rate is involved is true in that it is associated with responses to stress. Key structures are protected and cellular processes close down to a state of dormancy. Such stress responses, linked with reduced growth rate, will be driven at least in part by the high density and QS events occurring within the biofilm. It is not surprising that exceptional vegetative cell dormancy is the basic explanation of biofilm resistance. In biology, dormancy is a widespread survival response to stress.

REFERENCES

1. Adams, J. L., and McLean, R. J. C. 1999. Impact of *rpoS* deletion on *Escherichia coli* biofilms. Appl. Environ. Microbiol. **65**:4285–4287.
2. Al-Bakri, A. G., Gilbert, P., and Allison, D. G. 1999. Mixed species biofilms of *Burkholderia cepacia* and *Pseudomonas aeruginosa*. In J. Wimpenny, P. Gilbert,

J. Walker, M. Brading, and R. Bayston. (Eds.), *Biofilms: The Good, the Bad and the Ugly*, pp. 327–337. BioLine, Cardiff.

3. Allison, D., Maira-Litrán, T., and Gilbert, P. 1999. Perfused biofilm fermenters. Meth. Enzymol. **310**:232–248.

4. An, Y. H., Dickinson, R. B., and Doyle, R. J. 2000. Mechanisms of bacterial adhesion and pathogenesis of implant and tissue infections. In Y. H. An and R. J. Friedman (Eds.), *Handbook of Bacterial Adhesion: Principles, Methods, and Applications*, pp. 1–27. Humana Press Inc., Totowa, NJ.

5. Anderl, J. N., Franklin, M. J., and Stewart, P. S. 2000. Role of antibiotic penetration limitation in *Klebsiella pneumoniae* biofilm resistance to ampicillin and ciprofloxacin. Antimicrob. Agents Chemother. **44**:1818–1824.

6. Anwar, H., Brown, M. R. W., and Lambert, P. A. 1983. Effect of nutrient depletion on sensitivity of *Pseudomonas cepacia* to phagocytosis and serum bactericidal activity at different temperatures. J. Gen. Microbiol. **129**:2021–2027.

7. Bagge, N., Ciofu, O., Skovgaard, L. T., and Høiby, N. 2000. Rapid development in vitro and in vivo of resistance to ceftazidime in biofilm-growing *Pseudomonas aeruginosa* due to chromosomal beta-lactamase. APMIS **108**:589–600.

8. Boland, T., Latour, R. A., and Stutzenberger, F. J. 2000. Molecular basis of bacterial adhesion. In Y. H. An and R. J. Friedman. (Eds.), *Handbook of Bacterial Adhesion: Principles, Methods, and Applications*, pp. 29–41. Humana Press Inc., Totowa, NJ.

9. Brooun, A., Liu, S., and Lewis, K. 2000. A dose-response study of antibiotic resistance in *Pseudomonas aeruginosa* biofilms. Antimicrob. Agents Chemother. **44**:640–646.

10. Brown, M. R. W. 1977. Nutrient depletion and antibiotic susceptibility. J. Antimicrob. Chemother. **3**:198–201.

11. Brown, M. R. W., Allison, D., and Gilbert, P. 1988. Resistance of bacterial biofilms to antibiotics: A growth-rate related effect? J. Antimicrob. Chemother. **22**:777–780.

12. Brown, M. R. W., and Barker, J. 1999. Unexplored reservoirs of pathogenic bacteria: Protozoa and biofilms. Trends Microbiol. **7**:46–50.

13. Brown, M. R. W., Collier, P. J., Courcol, R. E., and Gilbert, P. 1995. Definition of phenotype in batch culture. In M. R. W. Brown and P. Gilbert (Eds.), *Microbiological Quality Assurance. A Guide Towards Relevance and Reproducibility of Inocula*, pp. 13–20. CRC Press, Boca Raton, FL.

14. Brown, M. R. W., Collier, P. J., and Gilbert, P. 1990. Influence of growth rate on susceptibility to antimicrobial agents: Modification of the cell envelope and batch and continuous culture studies. Antimicrob. Agents Chemother. **34**:1623–1628.

SMITH AND BROWN

15. Brown, M. R. W., and Williams, P. 1985. The influence of environment on envelope properties affecting survival of bacteria in infections. Annu. Rev. Microbiol. **39**:527–556.

16. Bühler, T., Ballestero, S., Desai, M., and Brown, M. R. W. 1998. Generation of a reproducible nutrient-depleted biofilm of *Escherichia coli* and *Burkholderia cepacia*. J. Appl. Bact. **85**:457–462.

17. Cashel, M., Gentry, D. R., Hernandez, V. J., and Vinella, D. 1996. The stringent response. In F. C. Neidhardt, R. Curtiss, III, J. L. Ingraham, E. C. C. Lin, K. B. Low, B. Magasanik, W. S. Reznikoff, M. Riley, M. Schaechter, and H. E. Umbarger (Eds.), Escherichia coli *and* Salmonella, pp. 1458–1496. ASM Press, Washington, D.C.

18. Chan, P. F., Foster, S. J., Ingham, E., and Clements, M. O. 1998. The *Staphylococcus aureus* alternative sigma factor sigma B controls the environmental stress response but not starvation survival or pathogenicity in a mouse abscess model. J. Bacteriol. **180**:6082–6089.

19. Clements, M. O., and Foster, S. J. 1999. Stress resistance in *Staphylococcus aureus*. Trends Microbiol. **7**:458–462.

20. Clewell, D. B. 1999. Sex pheromone systems in Enterococci. In G. M. Dunny and S. C. Winans (Eds.), *Cell–Cell Signaling in Bacteria*, pp. 47–66. ASM Press, Washington, D.C.

21. Costerton, J. W., Lewandowski, Z., Caldwell, D. E., Korber, D. R., and Lappin-Scott, H. M. 1995. Microbial biofilms. Annu. Rev. Microbiol. **49**:711–745.

22. Costerton, J. W., Stewart, P. S., and Greenberg, E. P. 1999. Bacterial biofilms: A common cause of persistent infections. Science **284**:1318–1322.

23. Cramton, S. E., Gerke, C., Nichols, W. W., and Gotz, F. 1999. The intercellular adhesion (*ica*) locus is present in *Staphylococcus aureus* and is required for biofilm. Infect. Immun. **67**:5427–5433.

24. Davies, D. G., Parsek, M. R., Pearson, J. P., Iglewski, B. H., Costerton, J. W., and Greenberg, E. P. 1998. The involvement of cell-to-cell signals in the development of a bacterial biofilm. Science **280**:295–298.

25. De Kievit, T. R., Gillis, R., Marx, S., Brown, C., and Iglewski, B. H. 2001. Quorum-sensing genes in *Pseudomonas aeruginosa* biofilms: Their role and expression patterns. Appl. Environ. Microbiol. **67**:1865–1873.

26. De Kievit, T. R., Parkins, M. D., Gillis, R. J., Srikumar, R., Ceri, H., Poole, K., Iglewski, B. H., and Storey, D. G. 2001. Multidrug efflux pumps: Expression patterns and contribution to antibiotic resistance in *Pseudomonas aeruginosa* biofilms. Antimicrob. Agents Chemother. **45**:1761–1770.

27. DeMaio, J., Zhang, Y., Ko, C., Young, D. B., and Bishai, W. R. 1996. A stationary-phase stress-response sigma factor from *Mycobacterium tuberculosis*. Proc. Natl. Acad. Sci. USA **93**:2790–2794.

28. Desai, M., Bühler, T., Weller, P. H., and Brown, M. R. W. 1998. Increasing resistance of planktonic and biofilm cultures of *Burkholderia cepacia* to ciprofloxacin and ceftazidime during exponential growth. J. Antimicrob. Chemother. **42**:153–160.

29. Dibdin, G., and Wimpenny, J. 1999. Steady-state biofilm: Practical and theoretical models. Meth. Enzymol. **310**:296–322.

30. Dodd, C. E. R., Sharman, R. L., Bloomfield, S. F., Booth, I. R., and Stewart, G. S. A. B. 1997. Inimical processes: Bacterial self destruction and sub-lethal injury. Trends Food Sci. Tech. **8**:238–241.

31. Duguid, I. G., Evans, E., Brown, M. R. W., and Gilbert, P. 1992. Growth-rate-independent killing by ciprofloxacin of biofilm-derived *Staphylococcus epidermidis*; evidence for cell-cycle dependency. J. Antimicrob. Chemother. **30**:791–802.

32. Duguid, I. G., Evans, E., Brown, M. R. W., and Gilbert, P. 1992. Effect of biofilm culture upon the susceptibility of *Staphylococcus epidermidis* to tobramycin. J. Antimicrob. Chemother. **30**:803–810.

33. Evans, D. J., Allison, D. G., Brown, M. R. W., and Gilbert, P. 1991. Susceptibility of *Pseudomonas aeruginosa* and *Escherichia coli* biofilms towards ciprofloxacin: Effect of specific growth rate. J. Antimicrob. Chemother. **27**:177–184.

34. Evans, K., Passador, L., Srikumar, R., Tsang, E., Nezezon, J., and Poole, K. 1998. Influence of the MexAB-OprM multidrug efflux system on quorum sensing in *Pseudomonas aeruginosa*. J. Bacteriol. **180**:5443–5447.

35. Fang, F. C., Libby, S. J., Buchmeier, N. A., Loewen, P. C., Switala, J., Harwood, J., and Guiney, D. G. 1992. The alternative σ factor KatF (RpoS) regulates *Salmonella* virulence. Proc. Natl. Acad. Sci. USA **89**:11,978–11,982.

36. Finch, J. E., and Brown, M. R. W. 1978. Effect of growth environment on *Pseudomonas aeruginosa* killing by rabbit polymorphonuclear leukocytes and cationic proteins. Infect. Immun. **20**:340–346.

37. Foley, I., Marsh, P., Wellington, E. M. H., Smith, A. W., and Brown, M. R. W. 1999. General stress response master regulator *rpoS* is expressed in human infection: A possible role in chronicity. J. Antimicrob. Chemother. **43**:164–165.

38. Gilbert, P., Allison, D. G., Evans, D. J., Handley, P. S., and Brown, M. R. W. 1989. Growth rate control of adherent bacterial populations. Appl. Environ. Microbiol. **55**:1308–1311.

39. Gilbert, P., Allison, D. G., Jacob, A., Korner, D., Wolfaa, G., and Foley, I. 1997. Immigration of planktonic *Enterococcus faecalis* cells into mature *E. faecalis* biofilms. In J. T. Wimpenny, P. Handley, P. Gilbert, and H. M. Lappin-Scott (Eds.), *Biofilms: Community Interactions and Control*, pp. 133–142. Bioline, Cardiff.

40. Gilbert, P., Coplan, F., and Brown, M. R. W. 1991. Centrifugation injury of gram-negative bacteria. J. Antimicrob. Chemother. 27:550–551.

41. Gilbert, P., Hodgson, A. E., and Brown, M. R. W. 1995. Influence of the environment on the properties of microorganisms grown in association with surfaces. In M. R. W. Brown and P. Gilbert (Eds.), *Microbiological Quality Assurance: A Guide Towards Relevance and Reproducibility of Inocula*, pp. 61–82. CRC Press, Boca Raton, FL.

42. Giwercman, B., Jensen, E. T., Høiby, N., Kharazmi, A., and Costerton, J. W. 1991. Induction of beta-lactamase production in *Pseudomonas aeruginosa* biofilm. Antimicrob. Agents Chemother. 35:1008–1010.

43. Gross, M., Cramton, S. E., Gotz, F., and Peschel, A. 2001. Key role of teichoic acid net charge in *Staphylococcus aureus* colonization of artificial surfaces. Infect. Immun. 69:3423–3426.

44. Hales, L. M., and Shuman, H. A. 1999. The *Legionella pneumophila rpoS* gene is required for growth within *Acanthamoeba castellani*. J. Bacteriol. 181:4879–4889.

45. Hassett, D. J., Ma, J. F., Elkins, J. G., McDermott, T. R., Ochsner, U. A., West, S. E. H., Huang, C. T., Fredericks, J., Burnett, S., Stewart, P. S., McFeters, G., Passador, L., and Iglewski, B. H. 1999. Quorum sensing in *Pseudomonas aeruginosa* controls expression of catalase and superoxide dismutase genes and mediates biofilm susceptibility to hydrogen peroxide. Mol. Microbiol. 34:1082–1093.

46. Havarstein, L. S., and Morrison, D. A. 1999. Quorum sensing and peptide pheromones in Streptococcal competence for genetic transformation. In G. M. Dunny and S. C. Winans (Eds.), *Cell–Cell Signaling in Bacteria*, pp. 9–26. ASM Press, Washington, D.C.

47. Hengge-Aronis, R. 1996. Regulation of gene expression during entry into stationary phase. In F. C. Neidhardt, R. Curtiss, III, J. K. Ingraham, E. C. C. Lin, K. B. Low, B. Magasanik, W. S. Reznikoff, M. Riley, M. Schaechter, and H. E. Umbarger (Eds.), Escherichia coli *and* Salmonella. *Cellular and Molecular Biology*, pp. 1497–1512. ASM Press, Washington, D.C.

48. Hengge-Aronis, R. 1999. Interplay of global regulators and cell physiology in the general stress response of *Escherichia coli*. Curr. Opin. Microbiol. 2:148–152.

49. Heys, S. J. D., Gilbert, P., Eberhard, A., and Allison, D. G. 1997. Homoserine lactones and bacterial biofilms. In J. Wimpenny, P. Handley, P. Gilbert, H. M. Lappin-Scott, and M. Jones (Eds.), *Biofilms: Community Interactions and Control*, pp. 103–112. Bioline, Cardiff.

50. Hopelman, A. I. M., and Tuomanen, E. 1992. Consequences of microbial attachment: Directing host cell functions with adhesins. Infect. Immun. 60:1729–1733.

51. Humphrey, T. J., Williams, A., McAlpine, K., Jorgensen, F., and O'Byrne, C. 1998. Pathogenicity in isolates of *Salmonella enterica* serotype Enteritidis PT4 which differ in RpoS expression: Effects of growth phase and low temperature. Epidemiol. Infect. **121**:295–301.

52. Jørgensen, F., Bally, M., Chapon-Herve, V., Stewart, G. S. A. B., Michel, G., Lazdunski, A., and Williams, P. 1999. RpoS-dependent stress tolerance in *Pseudomonas aeruginosa*. Microbiol. **145**:835–844.

53. Karthikeyan, S., Korber, D. R., Wolfaardt, G. M., and Caldwell, D. E. 2000. Monitoring the organization of microbial biofilm communities. In Y. H. An and R. J. Friedman (Eds.), *Handbook of Bacterial Adhesion: Principles, Methods, and Applications*, pp. 171–188. Humana Press, Totowa, NJ.

54. Kharazmi, A., Giwercman, B., and Høiby, N. 1999. Robbins device in biofilm research. Meth. Enzymol. **310**:207–215.

55. Krogfelt, K. A., Hjulgaard, M., Sorensen, K., Cohen, P. S., and Givskov, M. 2000. *rpoS* gene function is a disadvantage for *Escherichia coli* BJ4 during competitive colonization of the mouse large intestine. Infect. Immun. **68**:2518–2524.

56. Lazazzera, B. A., Palmer, T., Quisle, J., and Grossman, A. D. 1999. Cell-density control of gene expression and development in *Bacillus subtilis*. In G. M. Dunny and S. C. Winans (Eds.), *Cell–Cell Signaling in Bacteria*, pp. 27–46. ASM Press, Washington, D.C.

57. Lessard, I. A., and Walsh, C. T. 1999. VanX, a bacterial D-alanyl-D-alanine dipeptidase: Resistance, immunity, or survival function? Proc. Natl. Acad. Sci. USA **96**:11,028–11,032.

58. Lewis, K. 2001. Riddle of biofilm resistance. Antimicrob. Agents Chemother. **45**:999–1007.

59. Liedl, B. 2001. Catheter-associated biofilms. Curr. Opin. Urol. **11**:75–79.

60. Mah, T.-F. C., and O'Toole, G. A. 2001. Mechanisms of biofilm resistance to antimicrobial agents. Trends Microbiol. **9**:34–39.

61. Maira-Litrán, T., Allison, D. G., and Gilbert, P. 2000. Expression of the multiple antibiotic resistance operon(*mar*) during growth of *Escherichia coli* as a biofilm. J. Appl. Microbiol. **88**:243–247.

62. Matin, A. 1991. The molecular basis of carbon-starvation-induced general resistance in *Escherichia coli*. Mol. Microbiol. **5**:3–10.

63. McLeod, G. I., and Spector, M. P. 1996. Starvation- and stationary-phase-induced resistance to the antimicrobial peptide polymyxin B in *Salmonella typhimurium* is RpoS (sigma(S)) independent and occurs through both phoP-dependent and -independent pathways. J. Bacteriol. **178**:3683–3688.

64. Merrell, D. S., and Camilli, A. 1999. The *cadA* gene of *Vibrio cholerae* is induced during infection and plays a role in acid tolerance. Mol. Microbiol. **34**:836–849.

65. Moken, M. C., McMurray, L. M., and Levy, S. B. 1997. Selection of multiple-antibiotic-resistant (Mar) mutants of *Escherichia coli* by using the disinfectant pine oil: Roles of the *mar* and *acrAB* loci. Antimicrob. Agents Chemother. **41**:2770–2772.

66. Nichols, W. W. 1991. Biofilms, antibiotics and penetration. Rev. Med. Microbiol. **2**:177–181.

67. Nickerson, C. A., and Curtiss III, R. 1997. Role of sigma factor RpoS in initial stages of *Salmonella typhimurium* infection. Infect. Immun. **65**:1814–1823.

68. Novick, R. P. 1999. Regulation of pathogenicity in *Staphylococcus aureus* by a peptide-based density-sensing system. In G. M. Dunny and S. C. Winans (Eds.), *Cell–Cell Signaling in Bacteria*, pp. 129–146. ASM Press, Washington, D.C.

69. O'Toole, G. A., and Kolter, R. 1998. Flagella and twitching motility are necessary for *Pseudomonas aeruginosa* biofilm development. Mol. Microbiol. **30**:295–304.

70. Pearson, J. P., Van Delden, C., and Iglewski, B. H. 1999. Active efflux and diffusion are involved in transport of *Pseudomonas aeruginosa* cell-to-cell signals. J. Bacteriol. **181**:1203–1210.

71. Pier, G. B., Saunders, J. M., Ames, P., Edwards, M. S., Auerbach, H., Speert, D. P., and Hurwitch, S. 1987. Opsonophagocytic killing antibody to *Pseudomonas aeruginosa* mucoid exopolysaccharide in older noncolonized patients with cystic fibrosis. N. Engl. J. Med. **317**:793–798.

72. Pomposiello, P. J., and Demple, B. 2001. Redox-operated genetic switches: The SoxR and OxyR transcription factors. Trends Biotech. **19**:109–114.

73. Pratt, L. A., and Kolter, R. 1998. Genetic analysis of *Escherichia coli* biofilm formation: Roles of flagella, motility, chemotaxis and type I pili. Mol. Microbiol. **30**:285–293.

74. Public Health Laboratory Service. 2000. Surveillance of hospital-acquired bacteraemia in English hospitals 1997–1999. Public Health Laboratory Service, Colindale, London, UK.

75. Rachid, S., Ohlsen, K., Wallner, U., Hacker, J., Hecker, M., and Ziebuhr, W. 2000. Alternative transcription factor sigma B is involved in regulation of biofilm expression in a *Staphylococcus aureus* mucosal isolate. J. Bacteriol. **182**:6824–6826.

76. Rachid, S., Ohlsen, K., Witte, W., Hacker, J., and Ziebuhr, W. 2000. Effect of subinhibitory antibiotic concentrations on polysaccharide intercellular adhesin expression in biofilm-forming *Staphylococcus epidermidis*. Antimicrob. Agents Chemother. **44**:3357–3363.

77. Scott, J. M., Mitchell, T., and Haldenwang, W. G. 2000. Stress triggers a process that limits activation of the *Bacillus subtilis* stress transcription factor sigma B. J. Bacteriol. **182**:1452–1456.

78. Shiba, T., Tsutsumi, K., Yano, H., Iahara, Y., Yameda, A., Tanaka, T., Takahashia, M., Munekata, M., Rao, N. N., and Kornberg, A. 1997. Inorganic polyphosphate and the induction of *rpoS* expression. Proc. Natl. Acad. Sci. USA **94**:11,210–11,215.

79. Singh, P. K., Schaefer, A. L., Parsek, M. R., Moninger, T. O., Welsh, M. J., and Greenberg, E. P. 2000. Quorum-sensing signals indicate that cystic fibrosis lungs are infected with bacterial biofilms. Nature **407**:762–764.

80. Sissons, C. H., Wong, L. and An, Y. H. 2000. Laboratory culture and analysis of biofilms. In Y. H. An and R. J. Friedman (Eds.), *Handbook of Microbial Adhesion: Principles, Methods, and Applications*, pp. 133–169. Humana Press, Totowa, NJ.

81. Spector, M. P. 1998. The starvation-stress response (SSR) of Salmonella. Adv. Microb. Physiol. **40**:233–279.

82. Speranido, V., Mellies, J. L., Nguyen, W., Shin, S., and Kaper, J. B., 1999. Quorum sensing controls expression of the type III secretion gene transcription and protein secretion in enterohemorrhagic and enteropathogenic *Escherichia coli*. Proc. Natl. Acad. Sci. USA **96**:15,196–15,201.

83. Stewart, P. S. 1996. Theoretical aspects of antibiotic diffusion into microbial biofilms. Antimicrob. Agents Chemother. **40**:2517–2522.

84. Stewart, P. S., Roe, F., Rayner, J., Elkins, J. G., Lewandowski, Z., Ochsner, U. A., and Hassett, D. J. 2000. Effect of catalase on hydrogen peroxide penetration into *Pseudomonas aeruginosa* biofilms. Appl. Environ. Microbiol. **66**:836–838.

85. Stickler, D. J., Morris, N. S., McLean, R. J., and Fuqua, C. 1998. Biofilms on indwelling urethral catheters produce quorum-sensing signal molecules in situ and in vitro. Appl. Environ. Microbiol. **64**:3486–3490.

86. Surette, M. G., Miller, M. B., and Bassler, B. L. 1999. Quorum sensing in *Escherichia coli, Salmonella typhimurium*, and *Vibrio harveyi*: A new family of genes responsible for autoinducer production. Proc. Natl. Acad. Sci. USA **96**:1639–1644.

87. Sutherland, I. W. 2001. The biofilm matrix – an immobilised but dynamic environment. Trends Microbiol. **9**:222–227.

88. Sutton, M. D., Smith, B. T., Godoy, V. G., and Walker, G. C. 2000. The SOS response: Recent insights into umuDC-dependent mutagenesis and DNA damage tolerance. Annu. Rev. Genet. **34**:479–497.

89. Van Bambeke, F., Balzi, E., and Tulkens, P. M. 2000. Antibiotic efflux pumps. Biochem. Pharmacol. **60**:457–470.

90. Vuong, C., Saenz, H. L., Gotz, F., and Otto, M., 2000. Impact of the *agr* quorum-sensing system on adherence to polystyrene in *Staphylococcus aureus*. J. Infect. Dis. **182**:1688–1693.

SMITH AND BROWN

91. Whiteley, M., Parsek, M. R., and Greenberg, E. P. 2000. Regulation of quorum sensing by RpoS in *Pseudomonas aeruginosa*. J. Bacteriol. **182**:4356–4360.
92. Wimpenny, J., Manz, W., and Szewzyk, U. 2000. Heterogeneity in biofilms. FEMS Microbiol. Rev. **24**:661–671.
93. Wu, H., Song, Z., Givskov, M., Doring, G., Worlitzsch, D., Rygaard, J., and Høiby, N. 2001. *Pseudomonas aeruginosa* mutations in *lasI* and *rhlI* quorum sensing systems result in milder chronic lung infection. Microbiol. **147**:1105–1113.
94. Xu, K. D., McFeters, G., and Stewart, P. S. 2000. Biofilm resistance to antimicrobial agents. Microbiol. **146**:547–549.
95. Yasuda, H., Koga, T., and Fukoka, T. 1999. *In vitro* and *in vivo* models of bacterial biofilms. Meth. Enzymol. **310**:577–595.

CHAPTER 7

Tuberculosis

Yanmin Hu and Anthony R. M. Coates

INTRODUCTION

Tuberculosis (TB) is the single most important infectious disease in the world. This disease is caused by *Mycobacterium tuberculosis*. Although tuberculosis can be cured by antimicrobial therapy and can be prevented by a vaccine [Bacille Calmette-Guerin (BCG)], the total number of cases in the world is still increasing (52). There are 8 millions new cases every year, and over 3 million TB patients will die of this disease (33). After infection, which usually takes place in childhood, the bacteria are capable of persisting in a non-replicating or quiescent state for prolonged periods of time (73). These quiescent bacteria are called "immune persisters" by the authors, because they are induced by, and survive, the immune response (73). Reactivation of persisters, sometimes many years later, is responsible for most cases of active disease (59, 73). About one third of the world's population, two billion people, contain quiescent *M. tuberculosis* (60) and so this is by far the most common form of the microbe. These carriers provide a huge pool of potential disease, the lifetime risk of a carrier developing active tuberculosis being 5–10% (18). It is not known why many carriers break down into disease, but in some, weakening of the immune system, for example, by co-infection with HIV, is strongly associated with tuberculosis (18).

The main disease-control measures used for tuberculosis are case-finding and chemotherapy (3). However, chemotherapy itself induces persisters which are called "drug persisters" by the authors (30). These quiescent bacteria are very tolerant to the drug therapy and can only be killed slowly, if at all (30, 40, 41). This means that the period of chemotherapy is very long, about 6 months. Unfortunately, long periods of chemotherapy are not well tolerated by patients, who tend to stop taking the drugs before the end of the course, leading to disease relapse. Poor patient compliance is the single most important reason for the low efficacy of chemotherapy as a disease-control method.

THE QUIESCENT STATE OF *M. TUBERCULOSIS*

In order to survive within humans for long periods of time (decades in most cases), *M. tuberculosis* slows down its metabolic and growth rates and becomes more resistant to a variety of stress conditions (80) such as immune attack and antibiotics. This characteristic, the so-called quiescent state, is common to all bacteria (see Chapter 2) and is particularly well developed in *M. tuberculosis*. The slowing of the growth rate of *M. tuberculosis* needs to be distinguished from its inherent constitutional slow growth rate. The time required for *M. tuberculosis* to undergo one cycle of replication is 16–18 hours (72, 80), and for *M. leprae* it is about 12 days (80) under optimal conditions. By comparison, the doubling time of most other bacteria is 20 minutes during exponential growth. So, slowing the growth rate of *M. tuberculosis* is defined as a cycle of replication in excess of 18 hours, but the cycle is usually so slow that it cannot be readily detected.

As in other bacteria, one stress induces tolerance in *M. tuberculosis* to other stresses such as immune attack and antibiotics (see Chapter 2). It is likely that quiescent bacteria contain more than one population. For example, long before antibiotic chemotherapy was introduced, it was observed that bacteria which were visible in tuberculosis lesions by microscopy could not always be grown on solid media or by inoculation into animals (8). Cases are recorded in which thousand-fold-greater numbers of visible bacteria are seen than can be cultured. It is not known whether these culture-negative bacteria are a distinct subpopulation or are, simply, dead. However, it is clear that the population of quiescent bacteria which tolerates immune attack is different from, for example, the population which tolerates high doses of rifampicin. High-dose rifampicin kills most quiescent bacteria, and the antibiotic persister bacteria which remain cannot be cultured on solid media, in contrast to immune persisters which grow well on solid media (24, 30). So, the quiescent state contains several different populations, all of which, in the *M. tuberculosis*–infected patient, are tolerant to immune attack. Fig. 7.1 shows a hypothetical picture of subpopulations of tubercle bacilli within lesions, and this is discussed in more detail later in this chapter.

Role of the Immune Response

It has been shown in animal models that latent *M. tuberculosis* infection can be induced by the host's immune system. Sever and Youmans (54)

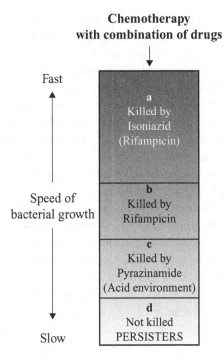

**Chemotherapy
with combination of drugs**

Fast

Speed of
bacterial growth

Slow

a
Killed by
Isoniazid
(Rifampicin)

b
Killed by
Rifampicin

c
Killed by
Pyrazinamide
(Acid environment)

d
Not killed
PERSISTERS

Figure 7.1. Hypothetical picture of sub-populations of tubercle bacilli within lesions
(adapted from 40).

observed that after intravenous infection of mice tubercle bacilli grew exponentially in the organs of the animals. After 3 weeks the bacteria stopped
growing. This cessation of growth was believed to be induced by the acquired
immune response. The immune response to *M. tuberculosis* is mediated by
T lymphocytes and macrophages (15, 23, 81). In 1893 Borrel and Metchnikoff
(4) first described that macrophages are involved in mycobacterial lung infection. The bacteria survive in macrophages, but their replication can be
suppressed by the influence of T lymphocytes directly and by the release of
cytokines (15, 23). In persons who have never been exposed to tubercle bacilli
the initial infection usually occurs in the lung (81), following the inhalation
of aerosols which contain *M. tuberculosis*. One or more bacilli settle within an
alveolus where they are rapidly engulfed by alveolar macrophages and seg-
mented neutrophils. The bacilli are highly resistant to destruction by host cells
and replicate within macrophages forming an initial lesion which is called a
Ghon focus (15, 23). Some bacteria are carried in macrophages to the hilar
lymph nodes in the chest where additional foci of infection develop. At the
same time, bacilli are spread further afield through the blood and the lymphat-
ics. This can lead to lesions in almost all of the organs and tissues of the body,

including other parts of the lungs. The majority of the disseminated tubercle bacilli are engulfed by mononuclear phagocytes in the organs and replicate almost as rapidly as they do at the site of the initial focus of infection in the lung. At this stage, the specific cell-mediated immune response has not been developed, but within about a week of infection, clones of antigen-specific T lymphocytes are produced (15, 23). These lymphocytes release lymphokines, which stimulate macrophages and cause them to form a compact cluster or granuloma around the foci of infection (15, 23). Some of these macrophages fuse to produce multinucleated giant cells. The centre of the granuloma contains a mixture of necrotic tissue and dead macrophages, which is referred to as caseation (15, 23). The activated macrophages are able to markedly inhibit the multiplication of virulent tubercle bacilli (15, 23). The formation of the granuloma is usually sufficient to arrest the primary infection, and surrounding fibroblasts produce dense scar tissue which walls off the lesion from the rest of the body. During the infection the host exhibits a delayed-type hypersensitivity reaction to certain mycobacterial proteins or polypeptides which can be demonstrated clinically by the tuberculin skin test (15, 23).

Most people who are infected with *M. tuberculosis* do not develop active tuberculosis. Rather, the bacteria become latent and survive in the organs for years in a quiescent form. There is an ongoing debate about the relative contribution of endogenous activation versus exogenous reinfection. Evidence shows that both are important. Post-primary tuberculosis is thought to be mainly due to activation of previously quiescent bacteria which have been lying latent in the patient for, often, long periods (51, 59). However, data from a study of an outbreak of multi-drug resistant tuberculosis in a shelter for the homeless showed that exogenous reinfection could occur in previously tuberculin-positive people (43). Phage typing and restriction-fragment-length polymorphism DNA fingerprinting have revealed high frequencies of exogenous reinfection in HIV-positive and -negative patients, and in immunocompetent tuberculin-positive patients (11, 34, 49, 55, 57, 62, 64).

When the immune system begins to fail, as occurs in AIDS and in older people, or is impaired by steroid therapy or malnutrition, latent tuberculosis tends to reactivate, resulting in an increasing number of *M. tuberculosis* in the organs, and clinical manifestations of tuberculosis (63). This situation has been studied in mice (7, 45). Orme (45) established a mouse model to study reactivation of latent infection which often occurs in the elderly. In this model, 3-month-old mice were infected with a sublethal aerogenic dose of *M. tuberculosis*. The tubercle bacilli multiplied rapidly in the lungs and spleens of infected mice for 1 to 3 months. Then the organisms stopped increasing in numbers, and the viable bacillary count remained constant until the mice were 18 months of age. The reason the number of organisms reaches a plateau is that the acquired host immune response attacks the bacteria, which respond by switching into the stress-resistant quiescent phase. At 18 months of age, when mice are into their old age, their T-cell immune response to tuberculosis declined, and the bacilli started to increase in number. Overall, in this experiment, the older the mice became, the greater the number of bacteria that were recorded in their organs. The mice died of overt tuberculosis at an average age of 24 months, 4 to 6 months earlier than the uninfected controls. A similar model has been established and shows that the immune system can induce and maintain M. tuberculosis in a persistent state (7). In this model, BCG-resistant and -susceptible mice were infected with a low dose (10^3 CFU) of M. tuberculosis. The numbers of the organisms increased and reached a peak in both strains of the mice between 35 and 43 days. Then the numbers of the bacilli decreased and latent infection was established after 65 days. Temporary suppression of the immune system of the mice by activation of the hypothalamic-pituitary-adrenal axis led to reactivation of the persistent M. tuberculosis in both strains of the mice.

(185)

TUBERCULOSIS

Quiescent Bacteria Associated with Antimicrobial Therapy

It has been proposed that during antimicrobial chemotherapy, four different populations of M. tuberculosis exist (40, 41). The majority of bacilli in the lesions of an untreated patient are actively dividing, probably at a rate similar to that in a log phase culture (Fig. 7.1, population a). These bacilli may be present in the lining of open cavities which have an abundant oxygen supply, and this favours the rapid growth of the bacteria. This population of microbes is rapidly killed by isoniazid, and to a lesser extent, by rifampicin. Organisms which are growing more slowly (Fig. 7.1, population b), for example in closed lesions where oxygen is limited, are not killed by isoniazid, but rifampicin is effective at killing slowly replicating bacteria. A further population (Fig. 7.1, population c) of slowly growing bacteria are thought to be located inside phagolysosomes within macrophages and are killed by pyrazinamide in this acid environment (40, 41). A small proportion of the bacilli persist in lesions in a non-replicating or quiescent state for an extended period of time (Fig. 7.1, population d). These quiescent bacilli tolerate the bactericidal effects of all known antituberculous drugs and are not killed by the host's immune system. They are termed "persisters." Even very high doses of rifampicin in vitro do not kill population d persisters (30).

The existence of persistent M. tuberculosis after chemotherapy in an animal model was shown for the first time by McCune and colleagues in the mid

1950's (35, 36). In this model, mice are infected intravenously with 10^5–10^6 colony forming units (CFU) of *M. tuberculosis*. Chemotherapy with isoniazid and pyrazinaminde is started immediately after the infection and continues for 12 weeks. After termination of the treatment, a "sterile" state is achieved which lasts for 4–6 weeks. At this stage, no tubercle bacilli are detected in homogenates of lungs or spleens of the mice by bacteriological culture or microscopy. Guinea pig inoculation of sterile mouse tissue also fails to show the presence of tubercle bacilli. This is a very sensitive animal inoculation technique. However, the disappearance of the bacilli does not mean that the organisms have been eliminated from the tissues. On the contrary, about 3 months after the end of drug treatment, cultivable bacilli are detected in about one-third of the mice. The reappearance of the tubercle bacilli in another two-thirds of the animals occurs 9 months after termination of the chemotherapy. Administration of a large dose of cortisone, which suppresses the immune system of the mice, accelerates the revival process of the tubercle bacilli. The persistence of the tubercle bacilli after 12 weeks of chemotherapy is not due to the emergence of drug-resistant mutants since the bacilli which are recovered from the mice remain essentially drug-sensitive (35). A series of similar studies was conducted later with a combination of isoniazid and rifampicin. Isoniazid and rifampicin in combination were given to *M. tuberculosis* infected mice for 12 months. A high relapse rate of 60% was observed by giving the mice 2 months of a high dose of cortisone (24). Further studies (10) suggest that even after 26 weeks of chemotherapy, virtually all nude mice relapse although no or very few relapses are seen in normal mice. This suggests that the immune system assists chemotherapy in its action against persistent bacteria. Although the original Cornell model and the later studies are unable to accurately mimic human tuberculosis, they go some way to show that tuberculosis can be converted from active disease to a latent infection with the use of chemotherapy. The persistent bacilli which survive the drug treatment can be reactivated and can cause active disease later.

Role of Oxygen

The progression of tuberculous disease, as shown by the rate of replication of virulent tubercle bacilli, is much slower in the liver and the spleen than it is in the lung. This is probably due to the oxygen tension, which is lower in these organs than in the alveoli of lung (81). Low levels of oxygen inhibit the growth of the bacteria (81). It has been proposed that an inadequate oxygen supply in closed tuberculosis lesions is the reason for the much smaller number of bacilli found in such lesions when compared with open cavities that are well supplied with air (8). In humans, oxygen availability is important in post-primary tuberculosis, which is a disease that occurs in individuals who have been previously infected by M. tuberculosis (81). In the immediate vicinity of bacteria within the lesions, caseous necrosis forms. This cheese-like material is due to the breakdown of cells and is located in the middle of the lesion, which is called a tuberculoma (23, 81). Bacteria survive in the tuberculoma, but replicate slowly, probably due in part to the limited oxygen supply in the lesion. Caseation is surrounded by granulomatous tissue containing activated macrophages, giant cells, lymphocytes and a varying degree of fibrosis (15, 23). The activated macrophages release proteases which cause softening and liquefaction of the caseous lesion. During the process of liquefaction, the tubercle bacilli multiply extracellularly for the first time and more rapidly than they do in the primary lesion, although still slowly. Eventually the expanding lesion erodes into a bronchus and the liquefied contents are discharged. At this point, air flows into the lesion and the bacteria multiply rapidly in the presence of oxygen. Some of the bacilli become detached from the cavity wall and enter the sputum. The patient is now highly infectious and transmits the bacteria via droplet spread to other contacts (15, 23).

Previous work (8, 48) indicates that oxygen tension plays an important role in the pathogenesis of tuberculosis, and suggests the possibility that low oxygen tension creates persistent bacilli. In rabbits and guinea pigs, reduction of atmospheric oxygen tension over a 3–4 week period inhibits the clinical progression of experimental tuberculosis (50). Virulence of M. tuberculosis, at least in part, depends on their ability to grow at a lower oxygen tension than their avirulent counterparts (53). In addition, M. tuberculosis produces enzymes of anaerobic metabolism in vivo and this could be due to the microaerophilic environment. This contrasts with the in vitro situation in which bacteria have lower levels of anaerobic enzymes because of the higher oxygen levels used to grow these organisms in culture (53).

Wayne (69–71) has developed an in vitro stationary-phase model. It is thought that the reduction in the rate of replication in this model is mainly due to the low oxygen tension in the cultures, but starvation may also be a factor. In the Wayne model, M. tuberculosis is grown in liquid culture without agitation. Initially bacilli grow at an exponential rate with a doubling time of 16–18 hours until the cell density reaches 10^7–10^8 CFU/ml. Bacteria fall to the bottom of the tube where dissolved oxygen becomes limiting. Such conditions are called microaerophilic and this results in a decrease in growth rate (68–70). Wayne examined the growth dynamics of M. tuberculosis in unagitated cultures by labelling the tubercle bacilli with ^{14}C and observing the distribution of the ^{14}C-labelled bacilli in the supernatant and in the deposit (69). The replication of the bacilli in the culture slowed down as oxygen was depleted as a result of self-consumption by the bacilli. At this time the logarithmic doubling in the upper layers of the culture balanced the rate at which the bacilli settled to the bottom of the container and created an oxygen gradient with low oxygen in the bottom of the tube (28). The organisms in the sediment adapt to microaerophilic conditions and enter a quiescent physiological state which lacks bacterial growth or replication. Wayne (70) demonstrated that on addition of [^{3}H]uracil to a culture of the quiescent state of *M. tuberculosis*, the bacilli immediately initiate RNA synthesis. However, the first DNA replication was not completed until 12 hours later. It is possible that during the period of settling the bacilli terminate their cell division before the DNA synthesis takes place. This means that non-replicating bacilli contain two copies of genomic DNA. The advantage of this genomic organisation is that the bacteria can resume growth immediately when conditions became favourable. Interestingly, Wayne also showed that the bacilli could divide synchronously when the organisms in the sediment are diluted with fresh medium and are incubated with continuous aeration. A further study conducted by Wayne and Sramek (74) confirmed that bacilli become quiescent under anaerobic conditions. They found that metronidazole, a drug that is specifically active against anaerobes, became bactericidal to stationary-phase *M. tuberculosis* under anaerobiosis. In contrast the stationary-phase bacteria exhibited tolerance to the bactericidal effects of rifampicin and isoniazid.

Gillespie *et al.* (20) examined phenotypic changes in mycobacteria, including *M. tuberculosis*, under anaerobic conditions. When the bacilli were shifted to anaerobic conditions, several changes occurred, including loss of acid-fastness, ability to grow on the malachite green-containing medium and iron-uptake activity. Shift of the cultures back to aerobic growth resulted in the return of all the characteristics which are usually associated with replicating mycobacteria. Cunningham and Spreadbury (14) studied the ultrastructural morphology of mycobacteria by transmission electron microscopy. They found that tubercle bacilli which were incubated in long-term microaerophilic or anaerobic conditions developed a very thickened cell wall outer layer, which may play an important role in preserving the cell structure during long-term stationary-phase survival.

During the course of adaptation to anaerobiosis, metabolic shutdown occurs accompanying a change from an oxygen-dependent pathway to the glyoxylate cycle (71). In Wayne's model, the slow-growing or non-replicating

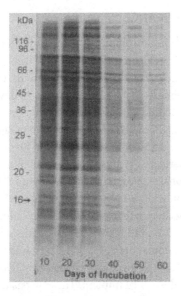

Figure 7.2. SDS-PAGE analysis of [^{35}S]methionine-labelled protein profiles of long-term stationary-phase cultures from 10 to 60 days. Each lane represents proteins extracted from 1.25×10^8 CFU bacilli. The results have been confirmed in two independent experiments.

bacilli expressed 10-fold higher levels of glycine dehydrogenase than the repli-cating organisms. Glycine dehydrogenase is an enzyme that catalyses the reductive amination of glyoxylate in association with oxidisation of NADH to NAD. The increasing glyoxylate synthesis may contribute a substrate for re-generation of NAD or ATP that might be needed for the bacilli to complete the final cycle of DNA replication before termination of their growth by anaer-obiosis. In further experiments, the authors of this chapter (26) measured protein synthesis of M. *tuberculosis* under microaerophilic and anaerobic con-ditions by [^{35}S]-methionine labelling and SDS-PAGE. They found that protein synthesis gradually decreases in long-term microaerophilic stationary-phase cultures (Fig. 7.2). Transfer of the cultures to strict anaerobic conditions re-sulted in a more rapid decrease and complete termination of detectable pro-tein synthesis. No changes in viable counts during prolonged microaerophilic incubation and under strict anaerobic conditions were observed within this experimental period (26). So, these data suggest that in unfavourable condi-tions, such as low oxygen, the bacteria adopt a strategy of metabolic shutdown. This increases their resistance to environmental challenges which helps to secure the viability of the cell during long-term existence in microaerophilic or anaerobic conditions.

ANTIBIOTIC TOLERANCE

Slow-growing or non-multiplying *M. tuberculosis* is tolerant to anti-tuberculous agents. Mitchison and Selkon (39) have shown that isoniazid and streptomycin are very active against log-phase bacilli, whereas isoniazid has no effect on stationary-phase cultures over a 14-day period, and strepto-mycin only causes a 10-fold decrease in viable counts over the same period. Later Dickinson *et al.* (17) observed that exposure of lag-phase (4 hours old) and log-phase (4 days) cultures to rifampicin for 40 to 160 minutes resulted in killing of the bacteria and a considerable reduction in uridine uptake. In contrast, stationary-phase 28-day cultures were not killed by rifampicin and uridine uptake was only reduced by a small margin (17).

The authors have shown that very old 100-day cultures which have been incubated under microaerophilic conditions contain at least two populations of bacteria, namely, population b (Fig. 7.1), which is sensitive to very high levels of rifampicin (100 μg/ml), and persisters (population d, Fig. 7.1), which are not killed by this dose of rifampicin. These persisters lose their ability to multiply on solid agar plates but can be resuscitated with liquid 7H9 medium. This subpopulation of stationary-phase bacilli is not killed by any known anti-tuberculous agents (30). RT-PCR analysis of RNA from persistent bacilli in the *in vitro* stationary-phase model showed that mRNA was still present in these organisms. These drug-tolerant persisters also incorporate [^{3}H]uridine into the RNA, which shows that ongoing transcription occurs in these organisms (30). These data show, for the first time, that persisters, at least in an *in vitro* model, are metabolically active.

LOCATION OF PERSISTERS *IN VIVO*

The location of non-multiplying *M. tuberculosis* persisters *in vivo* is un-known although they probably remain in solid caseous foci (15). Wayne and Salkin (66) examined resected human lung tissues and found that acid-fast bacilli were present in most of the old lesions from these tissues. These tu-bercle bacilli could be detected by microscopy but most of them could not be grown. About 20% of the closed lesions from these tuberculosis patients pro-duced small numbers of *M. tuberculosis* colonies. Surprisingly these patients converted to sputum negative more than 9 months before the resectional surgery was performed (66, 67). In another autopsy, almost no acid-fast bacilli were found by microscopy in the lung lesions of latent infection. However, guinea pig inoculation of homogenates of the lesions resulted in infection (44). Non-replicating bacteria were also found outside granulomas in the infected lungs (44). So persistent bacteria may be present in a non-acid-fast form in or outside granulomatous lesions.

PHYSIOLOGICAL STATE *IN VIVO*

The physiological state of non-multiplying *M. tuberculosis in vivo* is largely unknown. Whether the metabolic activities of persistent bacteria are completely switched off and whether they may remain metabolically active but at a level which is below the detection of the host immune system are unanswered questions (22, 80). Recent studies using the Cornell model have demonstrated that persistent bacteria are transcriptionally active (30). In this model, after 14 weeks of chemotherapy, all the mouse organs were apparently free of *M. tuberculosis* as determined by negative CFU counts and broth inoculation. However, mRNA was still detectable in the lungs of the mice by RT-PCR, suggesting that continuing metabolic activity is retained in the persistent bacilli. Messenger RNA is usually very short-lived, with a half-life of only a few minutes. A similar study has shown that mRNA for the 85 antigen is present *in vivo* in organs of infected mice (46). Further evidence from chemotherapy studies in humans suggests that persistent mycobacteria are metabolically active. For example, a 6–12 month course of chemotherapy with one bactericidal drug isoniazid reduced persistent bacilli in a high proportion of latently infected people (19). Isoniazid has very little sterilising effect against slow-growing *M. tuberculosis*. It is very unlikely that the bacilli whose metabolism is completely switched off would be susceptible to isoniazid.

THE MOLECULAR BASIS OF PERSISTENCE

The underlying molecular mechanisms by which *M. tuberculosis* can persist *in vitro* and *in vivo* for prolonged periods of time are currently under study in a number of groups in the world. So far, only limited attempts have been made to characterise gene expression associated with persistence of *M. tuberculosis* during latent infection and after chemotherapy. A major advancement in the genetics of mycobacteria has been the production of the complete genome sequence of *M. tuberculosis* (13). An integrated 4.4-megabase (Mb) map of a circular *M. tuberculosis* chromosome is now available (http://www.sanger.ac.uk/Projects/M_tuberculosis/), which provides a vast amount of new information about this important organism. It will open up many new directions for the investigation of *M. tuberculosis* persistence. The complete genome sequence of *M. tuberculosis* (13) has made the gene-wise analysis of the entire genome possible. In addition, modern microarray-based techniques enable quantitative analysis of the expression of many genes simultaneously, by determining changes in mRNA levels. DNA microarrays which contain most known or predicted open reading frames (ORFs) of the *M. tuberculosis* genome have been used to monitor drug-induced changes in gene expression (79) and to identify bacterial species (1). These results demonstrate the usefulness of DNA arrays of the entire genome to monitor differential gene expression. It will be very valuable to use microarray techniques to monitor differential gene expression at the mRNA level in stationary-phase and persistent bacteria. This may suggest specific drug targets against stationary-phase or persistent *M. tuberculosis*.

Global Gene Control

The capacity of bacteria to survive under unfavourable conditions is based on their ability to enter a specific programme of gene expression. Transcription is the initial event, which is mediated by an RNA polymerase. Global changes in gene expression are controlled by replacement of different sigma factors which play important roles in promoter recognition (32). The abundance of different sigma factors in a cell cycle is influenced by many factors, such as stationary phase and various stress conditions. The complete genomic sequence of *M. tuberculosis* shows two sigma 70 homologues, *sigA* and *sigB*, and a subfamily of the sigma 70 class called the extracytoplasmic function (ECF) family, which contains 11 alternative sigma factors with unknown physiological functions (*sigC, sigD, sigE, sigF, sigG, sigH, sigI, sigJ, sigK, sigL* and *sigM*) (13). Investigations into the role of each *M. tuberculosis* sigma factor during stationary-phase growth and under stress conditions will help us to understand the global control of gene expression in *M. tuberculosis*.

DeMaio *et al.* (16) have cloned a sigma factor gene, *sigF*, from *M. tuberculosis* using degenerate PCR. The deduced protein encoded by *M. tuberculosis sigF* shared significant homology to SigF sporulation sigma factors from *Streptomyces coelicolor* (47) and *Bacillus subtilis* (25) and to SigB, a stress-response sigma factor, from *B. subtilis* (25). Transcriptional analysis of *sigF* in *M. bovis* BCG by using an RNase protection assay revealed that *sigF* mRNA is significantly induced during stationary phase, under cold shock and nitrogen depletion, and weakly induced under other stress conditions such as oxidative stress, alcohol shock and microaerophilic stress and is absent during exponential growth. These findings indicate that the *M. tuberculosis sigF* gene encodes a stationary-phase and stress-response sigma factor which may be involved in response to environmental stress and in the adaptation of *M. tuberculosis* to persistent phase in the latent infection. Further genetic studies were conducted to construct a *sigF* gene mutated *M. tuberculosis* strain (12). The phenotype of the mutant was examined using a mouse model in which *M. tuberculosis* was intravenously injected. The mutated strain can grow as rapidly as the wild type during the first 8 weeks of infection. However, the mutant lost the ability to survive at the late stage of infection. This suggests that the *sigF* gene may be essential for *M. tuberculosis* to persist during latent infection. The *sigF* gene potentially directs a sporulation regulon (25, 47) but does not perform this role in *M. tuberculosis*, which is not a spore-forming bacterium, although *sigF* may play an analogous role in *M. tuberculosis* latency. Michele *et al.* (38) constructed an in-frame translational *lacZ-kan* fusion within the *sigF* gene to determine the conditions of *sigF* expression. The level of *sigF* reporter–specific expression was increased in a dose-dependent fashion after exposure of the bacilli to antibiotics such as ethambutol, rifampin, streptomycin and cycloserine. There was an over 100-fold induction of the level of *sigF* reporter–specific expression in late stationary-phase growth compared to that in exponential growth, confirming the finding of DeMaio *et al.* (16). Anaerobiosis induced *sigF* by greater than 150-fold, particularly in the presence of metronidazole. Cold shock increased the level of *sigF* specific expression, while heat shock decreased it. Oxidative stress also induced *sigF* specific expression. The induction of *sigF* after exposure to antibiotics suggests that this sigma factor may control genes that are important for mycobacterial persistence during chemotherapy.

Transcription of another sigma factor *sigB* gene was studied by the authors (27). They found an association between mRNA expression and low O_2 tension in bacteria that are entering stationary phase *in vitro*. Northern blot analysis showed that the mRNA level of the *sigB* gene was significantly induced during entry into stationary phase (Fig. 7.3) and under stress conditions (27). The pattern of the *sigB* transcription is very similar to that of the *sigF* gene in *M. tuberculosis* (16), the *rpoS* gene in *E. coli* (32, 61) and the *sigB* gene in *B. subtilis* (2, 5, 6). This finding suggests that SigB protein may be an alternative or secondary sigma factor that could control a large number of stationary-phase or stress regulons. It may play an important role in the ability of *M. tuberculosis* to survive in a quiescent form during latent tuberculosis infection.

In order to determine which genes are involved in maintaining viability of late-stationary-phase bacteria and persistent bacteria after antibiotic treatment, transcription of 82 genes of *M. tuberculosis*, including 13 genes which encode sigma factors, was examined by the authors using a mini-DNA array in the 100-day-old stationary-phase cultures before and after rifampicin treatment (31). The mRNA level of a sigma factor gene, *sigJ*, was strongly

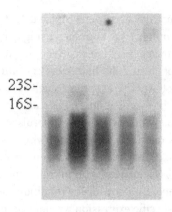

23S-
16S-

4 10 20 30 40
Days of Incubation

Figure 7.3. Northern blot analysis of steady-state levels of the *sigB* transcript. Total RNA was extracted from microaerophilic cultures of 4, 10, 20, 30 and 40 days and subjected to Northern blot analysis with the *sigB* gene–specific probes.

upregulated in the late stationary-phase cultures. Other genes including *sigI* were also upregulated, although to a lesser extent than *sigJ*. Surprisingly, there was no significant change in *sigJ* expression after rifampicin treatment, and most of the other 82 genes in the mini-DNA array also maintained expression. These results suggest that SigJ may control gene expression in the persistent state and may be an important component in the mechanisms by which *M. tuberculosis* survives prolonged stationary phase even in the presence of sterilising antibiotics. Construction of a *sigJ*-deleted mutant strain is under way in the same laboratory.

Specific Proteins

16-kDa protein: Yuan and colleagues (82) analysed the protein synthesis and profiles of *M. tuberculosis* during different growth phases using 2-D gels and ^{35}S-methionine labelling. They found that at least seven proteins were markedly induced during transition from log-phase to stationary-phase growth. One of these proteins was identified as the 16-kDa antigen which is an α-crystallin-like small heat shock protein (65). This protein is one of the most predominant of all the stationary phase–associated proteins. The gene encoding the 16-kDa protein, designated *hspX* (13), has been shown by Southern blot analysis to be restricted to the slow-growing *M. tuberculosis* complex of organisms. The protein was induced under microaerophilic

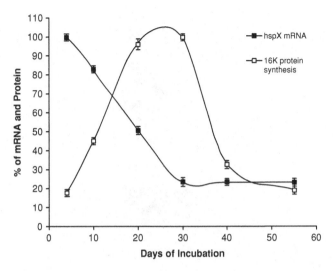

Figure 7.4. Relationship between the 16-kDa protein synthesis and the *hspX* mRNA accumulation during different growth phases under microaerophilic conditions. The densitometric data of the *hspX* mRNA were based on the signal of each mRNA normalised with the signal of 16S rRNA on the Northern blot. The densitometric data of the 16-kDa protein synthesis were based on the scanning data of the protein bands in the [^{35}S]methionine labelled SDS-PAGE.

conditions (82) by termination of aeration of a mid-log-phase culture, but not under other characterised environmental stresses. Overexpression of the 16-kDa protein in wild-type *M. tuberculosis* confers an enhanced resistance to autolysis on the bacilli and leads to a slower reduction in viability following the end of log-phase growth. This protein exhibits a significant ability to suppress the thermal aggregation of other proteins (9, 82), which suggests that it may act as a molecular chaperone. The α-crystallin-like 16-kDa protein may play an important role in long-term survival and persistence of *M. tuberculosis*. Transcription of the *hspX* gene has been studied by the authors using Northern blotting analysis with the RNA extracted from the cells grown in different growth phases (29). The cellular concentration of the *hspX* transcripts was maximal in the 4-day log-phase bacilli and then decreased as the cells reached stationary phase. In contrast, synthesis of the 16-kDa protein was low in 4-day cultures, increased as the bacteria entered stationary phase, and then fell away (see Fig. 7.4). The total amount of the 16-kDa protein increased with the enhanced protein synthesis but remained a constant level even when synthesis had returned to a low level. This probably reflects the highly stable nature of the 16-kDa protein. When transcriptional and translational levels

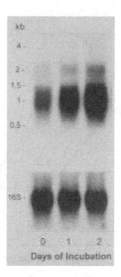

Figure 7.5. Northern blot analysis of the *hspX* mRNA of stationary-phase *M. tuberculosis* H37Rv after regrowth. RNA was extracted from a 30-day stationary-phase culture after incubation in fresh 7H9 broth for 0, 1 and 2 days and analysed by formaldehyde-agarose gel electrophoresis and Northern blotting. The filter was hybridized with an *hspX* gene–specific probe (top panel). The blot was stripped and reprobed with the 16S rRNA–specific probe (bottom panel). These results have been confirmed in two independent experiments.

of *hspX* gene expression are examined side by side (as shown in Fig. 7.4), it is clear that they are discordant, since in log phase there is a low level of the 16-kDa protein synthesis but a high level of *hspX* mRNA. In contrast, at 20–30 days, there is a high rate of protein synthesis but a low amount of mRNA. This suggests that an increase in protein synthesis is accompanied by a decrease in mRNA accumulation during the transition from log phase to stationary phase under microaerophilic conditions.

The inverse relationship between translation and transcription is reversible because, when the bacilli are shifted from stationary phase back to exponential growth, a rapid increase in the level of the *hspX* mRNA is observed (see Fig. 7.5).

A similar situation is seen in anaerobic conditions. The authors have shown that, when the bacilli are incubated anaerobically, transcription initiation of *hspX* rapidly decreases (see Fig. 7.6). However, when the bacteria are shifted from anaerobiosis back to aerobic conditions, *hspX* mRNA rises back to its usual level in log-phase cultures (Fig. 7.6) and this is accompanied by

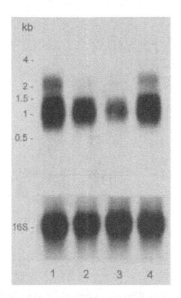

Figure 7.6. Northern blot analysis of the *hspX* mRNA from *M. tuberculosis* H37Rv under anaerobic and aerobic conditions. RNA was extracted from a 30-day stationary-phase culture under anaerobic conditions for 0, 1 and 4 hours following 12 hours of aerobic incubation. Lanes 1, 2 and 3: anaerobic incubation for 0, 1, and 4 hours. Lane 4: 12 hours aerobic incubation after 4 hours of anaerobiosis. The filter was hybridized with an *hspX* gene–specific probe (top panel). The blot was stripped and reprobed with a 16S rRNA–specific probe (bottom panel). These results have been confirmed in two additional experiments.

only a small increase in 16-kDa protein synthesis (Fig. 7.7). These data show that *hspX* gene expression during entry into the stationary phase is regulated by a post-transcriptional control mechanism because the synthesis of mRNA and protein is uncoupled.

Isocitrate lyase: The *icl* gene encodes an enzyme, isocitrate lysae, which is involved in the glyoxylate shunt. McKinney and colleagues examined the role of the *icl* gene during dormancy of *M. tuberculosis* in the mice (37). They constructed an *icl* gene–disrupted mutant and infected mice with both mutant and the wild type. They found that during the first 2 weeks of infection, the *icl* mutant grew as well as the wild-type strain. However, after the initial 2-week period, the CFU counts of the mutant fell dramatically, whilst the viability of the wild type remained virtually unchanged. These data indicate that the mutant cells lose the ability to survive in the late stages of infection where the organisms are slowly replicating or are not growing at all. Since isocitrate

(198)

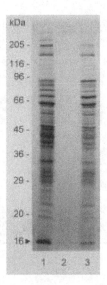

Figure 7.7. SDS-PAGE analysis of proteins synthesised by a 30-day stationary-phase culture after 1-week of anaerobic incubation with and without reintroduction of O_2. Protein synthesis was examined by [^{35}S]methionine labelling, SDS-PAGE and fluorography. Lanes: 1, microaerophilic control; 2, 1-week anaerobiosis; 3, oxygen introduced after 1 week of anaerobic incubation. Arrows indicate the 16-kDa protein.

lyase is an enzyme in the glyoxylate shunt, the mutant-derived data suggest that *M. tuberculosis* needs this enzyme in order to convert fatty acids to sugar precursors which provide energy for the bacteria to survive in a persistent form in latent infection. The key role which isocitrate lyase seems to play in *M. tuberculosis* persistence makes it an attractive target for drug development. Protein inhibitors based on its structure, which has recently been solved (56), may lead to potent agents which are more effective against slow-growing or non-multiplying bacteria.

TUBERCULOSIS DISEASE CONTROL

Currently, the main method of tuberculosis disease control is case-finding and chemotherapy. Clearly, at a global level where the disease is now epidemic, there are unresolved problems. These problems stem from the inability of antimicrobial agents or the immune system to eradicate non-replicating or quiescent *M. tuberculosis*. As a result of the persistent nature of these bacteria, a long period of chemotherapy, 6 months, is needed to treat tuberculosis. Such extended periods of treatment inevitably lead to major problems in compliance (58). Poor compliance means that patients stop taking their drugs before the end of the course of therapy which results in disease relapse, sometimes with multi-drug-resistant strains (42). Non-compliant patients who relapse, then infect further contacts (42).

In order to interrupt the cycle of long periods of chemotherapy, the high incidence of poor compliance, the relapse of non-compliant patients and the infection of contacts, it is necessary to shorten the period of chemotherapy. This could be achieved, theoretically, if persistent *M. tuberculosis* which survives chemotherapy, could be killed. In this situation, chemotherapy regimens could be shortened and so compliance would be improved. In addition, a shorter period of chemotherapy should result in a lower incidence of drug-resistant strains which tend to emerge in non-compliant patients. So, an ideal anti-tuberculosis drug should, firstly, rapidly kill large numbers of actively metabolising or replicating bacilli in order to prevent their subsequent multiplication during and after the completion of treatment. Secondly, it must exhibit a sterilising effect on slowly or intermittently metabolising bacilli. Thirdly, it must prevent the emergence of drug resistance (21).

The problem with current anti-tuberculous drugs, namely a combination of isoniazid, rifampicin, ethambutol and pyraziamide or other drugs in 6-month chemotherapy, lies in their inability to accomplish a fast kill of quiescent bacteria, although they do eventually reduce the numbers of persisters. In contrast, these drugs kill actively replicating bacteria relatively quickly. Isoniazid is bactericidal and is mainly responsible for killing the rapidly growing bacilli in the lesion during the first few days of treatment, but it has a low sterilising activity (see Fig. 7.1 (a))(40, 41). Rifampicin is not only bactericidal but more also, and importantly, exhibits a sterilising activity (Fig. 7.1 (b)) (40, 41). Rifampicin starts killing *M. tuberculosis* in culture within 1 hour of exposure to the drug, in contrast to isoniazid, which only starts killing after at least 24 hours (40, 41). Rifampicin can effectively kill the bacilli which have short spurts of active metabolism (40, 41). Pyrazinamide has very weak early bactericidal activity but is a potent sterilizing drug at acid pH, which kills the bacilli whose metabolism is inhibited by an acid environment (Fig. 7.1 (c)) (40, 41). At present, there is no drug which will kill persistent bacilli (Fig. 7.1 (d)). Two months of treatment of tuberculosis patients with the combination of the four drugs can effectively arrest the active infection and stop the transmission of the disease. Another 4 months of continuing treatment is needed in order prevent a high proportion of relapses. Presumably, persistent bacteria do not remain in an antibiotic-tolerant state forever but rather have spurts of active metabolism from time to time which renders them susceptible at some point during the last 4 months of chemotherapy. So, completion of 6 months of treatment with a combination of three or four drugs is essential for the successful chemotherapy of tuberculosis. However, in practice, such a long period of chemotherapy is bound to present many difficulties. The most common problem is that, after a few months of treatment, the patients feel better and stop taking their drugs (75). This failure of compliance leads to relapse of tuberculosis. In addition, early discontinuation of chemotherapy will introduce genetic multi-drug resistance.

In order to address the problem of poor compliance during chemotherapy, the World Health Organisation (WHO) has established a policy of full supervision of chemotherapy [directly observed treatment (DOT)] for every tuberculosis patient (76). DOT requires doctors to diagnose tuberculosis cases and supervise the administration of each and every dose of drugs to patients over a period of 6 to 8 months. The introduction of DOT has increased compliance in those countries which have successfully introduced this regimen (77). However, DOT needs sufficient infrastructure and resources to deliver an experienced health care team, which are deficient in many developing countries where most tuberculosis cases and deaths occur (78).

A separate but related issue is the relapse rate in previously healthy individuals. Of the two billion people who are infected with *M. tuberculosis*, most harbour the bacterium from childhood and carry it for the rest of their lives. These individuals have a 5–10% lifetime rate of active tuberculosis. There is no vaccine or drug-therapeutic regimen which is currently in use that will prevent tuberculosis in this huge pool of people. Therefore, new drugs are required to eliminate the slow-growing and non-replicating bacteria which survive the host immune system in carriers. These new drugs will need to not only kill bacteria in latent, healthy carriers, but also shorten the duration of treatment and be effective against multi-drug-resistant *M. tuberculosis*. Better understanding of the molecular mechanisms of *M. tuberculosis* persistence is particularly important in devising efficient disease-control policies and improvements in chemotherapy, including designing new drugs with a direct toxic effect on persisters or new compounds that can inhibit essential steps in entry of the bacteria into the persistent state.

CONCLUSION

The elimination of tuberculosis as a global disease is not feasible at the moment because of the formation of the quiescent *M. tuberculosis* state in the human body. This means that these bacteria can survive immune defences and chemotherapy. In the future, studies of differential gene expression at the transcriptional and protein levels in *in vitro* and *in vivo* models will help us to discover the essential genes, which may be involved in *M. tuberculosis* persistence. Subsequent knock-out of these genes should establish the specific functions of these genes in latency of *M. tuberculosis* in animals. These strategies will provide drug targets for the development of anti-tuberculosis agents against-slow-growing or non-multiplying organisms and may also provide attenuated strains for more effective vaccines.

REFERENCES

1. Behr, M. A., Wilson, M. A., Gill, W. P., Salamon, H., Schoolnik, G. K., Rane, S., and Small, P. M. 1999. Comparative genomics of BCG vaccines by whole-genome DNA microarray. Science **284**:1520–1523.

2. Benson, A. K., and Haldenwang, W. G. 1993. The σ^B-dependent promoter of the *Bacillus subtilis sigB* operon is induced by heat shock. J. Bacteriol. **175**:1929–1935.

3. Borgdorff, M. W., Floyd, K., and Broekmans, J. F. 2002. Interventions to reduce tuberculosis mortality and transmission in low- and middle-income countries. Bull. World Health Org. **80**:217–227.

4. Borrel, A., and Metchnikoff, M. 1893. Tuberculose pulmonaire experimentale. Ann. Inst. Pasteur **8**:594–625.

5. Boylan, S. A., Redfield, A. R., and Price, C. W. 1993a. Transcription factor σ^B of *Bacillus subtilis* controls a large stationary-phase regulon. J. Bacteriol. **175**:3957–3963.

6. Boylan, S. A., Redfield, A. R., Brody, M. S., and Price, C. W. 1993b. Stress-induced activation of the σ^B transcription factor of *Bacillus subtilis*. J. Bacteriol. **175**:7931–7937.

7. Brown, D. H., Miles, B. A., and Zwilling, B. S. 1995. Growth of *Mycobacterium tuberculosis* in BCG-resistant and -susceptible mice: Establishment of latency and reactivation. Infect. Immun. **63**:2243–2247.

8. Canetti, G. J. 1955. The tubercle bacillus in the pulmonary lesion of man. In *The Histobacteriogenesis of Tuberculous Lesions: Experimental Studies*, pp. 87–90. Spring Publishing Company, New York.

9. Chang, Z., Primm, T. P., Jakana, J., Lee, I. H., Serysheva, I., Chiu, W., Gilbart, H. F., and Quiocho, F. A. 1996. *Mycobacterium tuberculosis* 16kDa antigen (Hsp16.3) functions as an oligomeric structure *in vitro* to suppress thermal aggregation. J. Biol. Chem. **271**:7218–7223.

10. Chapuis, L., Ji, B., Truffot-Pernot, C., O'Brien, R. J., Raviglione, M. C., and Grosset, J. H. 1994. Preventive therapy of tuberculosis with rifapentine in immunocompetent and nude mice. Am. J. Respir. Crit. Care. Med. **150**:1355–1362.

11. Chaves, F., Dronda, F., Alonso-Sanz, M., and Noriega, A. R. 1999. Evidence of exogenous reinfection and mixed infection with more than one strain of *Mycobacterium tuberculosis* among Spanish HIV-infected inmates. AIDS 13:615–620.

12. Chen, P., Ruiz, R. E., Li, Q., Silver, R. F., and Bishai, W. R. 2000. Construction and characterization of a *Mycobacterium tuberculosis* mutant lacking the alternate sigma factor gene, sigF. Infect. Immun. 68:5575–5580.

13. Cole, S. T., Brosch, R., Parkhill, J., Garnier, T., Churcher, C., Harris, D., Gordon, S. V., Eiglmeier, K., Gas, S., Barry, C. E., III, Tekaia, F., Badcock, K., Basham, D., Brown, D., Chillingworth, T., Connor, R., Davies, R., Devlin, K., Feltwell, T., Gentles, S., Hamlin, N., Holroyd, S., Hornsby, T., Jagels, K., Krogh, A., McLean, J., Moule, S., Murphy, L., Oliver, K., Osborne, J., Rajandream, M.-A., Rogers, J., Rutter, S., Seeger, K., Skelton, J., Squares, R., Squares, S., Sulston, J. E., Taylor, K., Whitehead, S., and Barrell, B. G. 1998. Deciphering the biology of *Mycobacterium tuberculosis* from the complete genome sequence. Nature 393:537–544.

14. Cunningham, A. F., and Spreadbury, C. L. 1998. Mycobacterial stationary phase induced by low oxygen tension: Cell wall thickening and locatization of the 16-kilodalton α-crystallin homolog. J. Bacteriol. 180:801–808.

15. Dannebery, A. M., and Rook, G. A. W. 1994. Pathogenesis of pulmonary tuberculosis: An interplay of tissue-damaging and macrophage-activating immune responses-dual mechanisms that control bacillary multiplication. In *Tuberculosis: Pathogensis, Protection, and Control*, pp. 459–501 (Bloom, B. R., Ed.). American Society for Microbiology, Washington, D.C.

16. DeMaio, J., Zhang, Y., Ko, C., Young, D. B., and Bishai, W. R. 1996. A stationary-phase stress-response sigma factor from *Mycobacterium tuberculosis*. Proc. Natl. Acad. Sci. USA 93:2790–2794.

17. Dickinson, J. M., Jackett, P. S., and Mitchison, D. A. 1972. The effect of pulsed exposures to rifampicin on the uptake of uridine-C by *Mycobacterium tuberculosis*. Am. Rev. Respir. Dis. 105:519.

18. Dolin, P. J., Raviglione, M. C., and Kochi, A. 1994. Global tuberculosis incidence and mortality during 1990–2000. Bull. World Health Org. 72:213–220.

19. Farer, L. S. 1982. Chemoprophylaxis. Am. Rev. Respir. Dis. 125(suppl.):102–107.

20. Gillespie, J., Barton, L. L., and Rypka, E. W. 1986. Phenotypic changes in mycobacteria grown in oxygen-limited conditions. J. Med. Microbiol. 21:251–255.

21. Girling, D. J. 1989. The chemotherapy of tuberculosis. In *Clinical Aspects of*

Mycobacterial Disease, pp. 285–323, Ratledge, C., Stanford, J., and Grange, J. M., Eds. Academic Press, London, San Diego.

22. Grange, J. M. 1992. The mystery of the mycobacterial 'persistor.' Tuber. Lung Dis. **73**:249–251.

23. Grange, J. M. 1998. Pathogenesis of mycobacterial disease. In *Mycobacteria. I Basic Aspects*, pp. 145–177 (Gangadharam, P. R. J., and Jenkins, P. A., Eds.). Chapman and Hall, New York.

24. Grosset, J. 1978. The sterilizing value of rifampicin and pyrazinamide in experimental short course chemotherapy. *Tubercle* **59**:287–297.

25. Haldenwang, W. G. 1995. The sigma factor of *Bacillus subtilis*. Microbiol. Rev. **59**:1–30.

26. Hu, Y. M., Butcher, P. D., Sole, K., Mitchison, D. A., and Coates, A. R. M. 1998. Protein synthesis is shutdown in dormant *Mycobacterium tuberculosis* and is reversed by oxygen or heat shock. FEMS Microbiol. Lett. **158**:139–145.

27. Hu, Y., and Coates, A. R. M. 1999. Transcription of two sigma 70 homologue genes, *sigA* and *sigB*, in stationary-phase *Mycobacterium tuberculosis*. J. Bacteriol. **181**:469–476.

28. Hu, Y., Butcher, P. D., Mangan, J. A., Rajandream, M.-A., and Coates, A. R. M. 1999. Regulation of *hmp* gene transcription in *Mycobacterium tuberculosis*: Effects of oxygen limitation and nitrosative and oxidative stress. J. Bacteriol. **181**:3486–3493.

29. Hu, Y., and Coates, A. R. M. 1999. Transcription of the stationary-phase-associated *hspX* gene of *Mycobacterium tuberculosis* is inversely related to synthesis of the 16-kilodalton protein. J. Bacteriol. **181**:1380–1387.

30. Hu, Y., Mangan, J. A., Dhillon, J., Sole, K. M., Mitchison, D. A., Butcher, P. D., and Coates, A. R. M. 2000. Detection of mRNA transcripts and active transcription in antibiotic-induced models of *Mycobacterium tuberculosis* persistence. J. Bacteriol. **182**:6358–6365.

31. Hu, Y., and Coates, A. R. M. 2001. Increased levels of *sigJ* mRNA in late stationary phase cultures of *Mycobacterium tuberculosis* detected by DNA array hybridisation. FEMS Microbiol. Lett. **202**:59–65.

32. Jishage, M., and Ishihama, A. 1995. Regulation of RNA polymerase sigma subunit synthesis in *Escherichia coli*: Intracellular levels of σ^{70} and σ^{38}. J. Bacteriol. **177**:6832–6835.

33. Kochi, A. 1991. The global tuberculosis situation and the new control strategy of the World Health Organization. *Tubercle* **72**:1–6.

34. Le, H. Q., and Davidson, P. T. 1996. Reactivation and exogenous reinfection: Their relative roles in the pathogenesis of tuberculosis. *Curr. Clin. Top. Infect. Dis.* **16**:260–276.

35. McCune, R. M., Tompsett, R., and McDermott, W. 1956. The fate of *Mycobacterium tuberculosis* in mouse tissues as determined by the microbial enumeration technique II. The conversion of tuberculous infection to the latent state by administration of pyrazinamide and a companion drug. J. Exp. Med. **104**:763–802.

36. McCune, R. M., Feldmann, F. M., Lambert, H. P., and McDermott, W. 1966. Microbial persistence I. The capacity of tubercle bacilli to survive sterilization in mouse tissues. J. Exp. Med. **123**:445–468.

37. McKinney, J. D., Honer zu Bentrup, K., Munoz-Elias, E. J., Miczak, A., Chen, B., Chan, W. T., Swenson, D., Sacchettini, J. C., Jacobs, W. R., Jr., and Russell, D. G. 2000. Persistence of *Mycobacterium tuberculosis* in macrophages and mice requires the glyoxylate shunt enzyme isocitrate lyase. Nature **406**:735–738.

38. Michele, T. M., Ko, C., and Bishai, W. R. 1999. Exposure to antibiotics induces expression of the *Mycobacterium tuberculosis sigF* gene: Implications for chemotherapy against mycobacterial persistors. Antimicrob. Agents Chemother. **43**:218–225.

39. Mitchison, D. A., and Selkon, J. B. 1956. The bactericidal activities of antituberculous drugs. Am. Rev. Tuberc. **74**(supp.):109–116.

40. Mitchison, D. A. 1979. Basic mechanisms of chemotherapy. Chest **76**:771–781.

41. Mitchison, D. A. 1997. Mechanisms of tuberculosis chemotherapy. J. Pharm. Pharmacol. **49**(supp.1):31–36.

42. Mitchison, D. A. 1998. How drug resistance emerges as a result of poor compliance during short course chemotherapy for tuberculosis. Int. J. Tuberc. Lung Dis. **2**:10–15.

43. Nardell, E., McInnis, B., Thomas, B., and Weidhaas, S. 1986. Exogenous reinfection with tuberculosis in a shelter for the homeless. N. Engl. J. Med. **315**:1570–1575.

44. Opie, E. L., and Aronson, J. D. 1927. Tubercle bacilli in latent tuberculous lesions and in lung tissue without tuberculous lesions. Arch. Pathol. Lab. Med. **4**:1–21.

45. Orme, I. M. 1988. A mouse model of the recrudescence of latent tuberculosis in the elderly. Am. Rev. Respir. Dis. **137**:716–718.

46. Pai, S. R., Actor, J. K., Sepulveda, E., Hunter, R. L., Jr., and Jagannath, C. 2000. Identification of viable and non-viable *Mycobacterium tuberculosis* in mouse organs by directed RT-PCR for antigen 85B mRNA. Microb. Pathog. **28**:335–342.

47. Potuckova, L., et al. 1995. A new RNA polymerase sigma factor, sigma F, is required for the late stages of morphological differentiation in Streptomyces spp. Mol. Microbiol. **17**:37–48.

48. Raffel, S. 1956. Immunopathology of tuberculosis. Am. Rev. Tuberc. **Aug. Suppl.**: 60–74.

49. Raleigh, J. W., and Wichelhausen, R. 1973. Exogenous reinfection with *Mycobacterium tuberculosis* confirmed by phage typing. Am. Rev. Respir. Dis. **108**:639–642.

50. Rich, A. R., and Follis, R. H., Jr. 1942. The effect of low oxygen tension upon the development of experimental tuberculosis. Bull. Johns Hopkins Hospital **71**:345.

51. Romeyn, J. A. 1970. Exogenous reinfection in tuberculosis. Am. Rev. Respir. Dis. **101**:923–927.

52. Schweinle, J. E. 1990. Evolving concepts of the epidemiology diagnosis, and therapy of *Mycobacterium tuberculosis* infection. Yale J. Biol. Med. **117**:191–196.

53. Segal, W. 1984. Growth dynamics of in vivo and in vitro grown mycobacterial pathogens. In *The Mycobacteria – A Sourcebook*, pp. 547–573 (Kubica, G. P., and Wayne, L. G., Eds.). Marcel Dekker, New York.

54. Sever, J. L., and Youmans, G. P. 1957. Enumeration of viable tubercle bacilli from the organs of nonimmunized and immunized mice. Am. Rev. Tuberc. Pulmon. Dis. **76**:616–635.

55. Shafer, R. W., Singh, S. P., Larkin, C., and Small, P. M. 1995. Exogenous reinfection with multidrug-resistant *Mycobacterium tuberculosis* in an immunocompetent patient. Tuberc. Lung Dis. **76**:575–577.

56. Sharma V., Sharma, S., Hoener zu Bentrup, K., McKinney, J. D., Russell, D. G., Jacobs, W. R., Jr., and Sacchettini, J. C. 2000. Structure of isocitrate lyase, a persistence factor of Mycobacterium tuberculosis. Nat. Struct. Biol. **7**:663–668.

57. Small, P. M., Shafer, R. W., Hopewell, P. C., Singh, S. P., Murphy, M. J., Desmond, E., Sierra, M. F., and Schoolnik, G. K. 1993. Exogenous reinfection with multidrug-resistant *Mycobacterium tuberculosis* in patients with advanced HIV infection. N. Engl. J. Med. **328**:1137–1144.

58. Stanford, J. L., Grange, J. M., and Pozniak, A. 1991. Is Africa lost? Lancet **338**:557–558.

59. Stead, W. W. 1967. Pathogenesis of a first episode of chronic pulmonary tuberculosis in man: Recrudescence of residuals of the primary infection or exogenous reinfection? Am. Rev. Respir. Dis. **95**:729–745.

60. Sudre, P., ten Dam, G., and Kochi, A. 1992. Tuberculosis: A global overview of the situation today. Bull. World Health Org. **70**:149–159.

61. Tanaka, K., Takayanagi, Y., Fujita, N., Ishihama, A., and Takahashi, H. 1993. Heterogeneity of the principal σ factor in *Escherichia coli*: The *rpoS* gene product, σ^{38}, is a second principal σ factor of RNA polymerase

in stationary-phase *Escherichia coli*. Proc. Natl. Acad. Sci. USA **90**:3511–3515.

62. Turett, G. S., Fazal, B. A., Justman, J. E., Alland, D., Duncalf, R. M., and Telzak, E. E. 1997. Exogenous reinfection with multidrug-resistant *Mycobacterium tuberculosis*. Clin. Infect. Dis. **24**:513–514.

63. Ulrichs, T., and Kaufmann, S. H. 2002. Mycobacterial persistence and immunity. Front Biosci. **7**:458–469.

64. Van Rie, A., Warren, R., Richardson, M., Victor, T. C., Gie, R. P., Enarson, D. A., Beyers, N., and Van Helden, P. D. 1999. Exogenous reinfection as a cause of recurrent tuberculosis after curative treatment. N. Engl. J. Med. **341**:1174–1179.

65. Verbon, A., Hartskeerl, R. A., Schuitema, A., Kolk, A. H., Young, D. B., and Lathigra, R. 1992. The 14,000-molecular-weight antigen of *Mycobacterium tuberculosis* is related to the alpha crystallin family of low-molecular-weight heat shock proteins. J. Bacteriol. **174**:1352–1359.

66. Wayne, L. G., and Salkin, D. 1956. The bacteriology of resected tuberculous pulmonary lesions. I. The effect of interval between reversal of infectiousness and subsequent surgery. Am. Rev. Tubercu. Pulmon. Dis. **74**:376–387.

67. Wayne, L. G. 1960. The bacteriology of resected tuberculous pulmonary lesions. II. Observation on bacilli which are stainable but which cannot be cultured. Am. Rev. Respir. Dis. **82**:370–377.

68. Wayne, L. G., and Diaz, G. A. 1967. Autolysis and secondary growth of *Mycobacterium tuberculosis* in submerged culture. J. Bacteriol. **93**:1374–1381.

69. Wayne, L. G. 1976. Dynamics of submerged growth of *Mycobacterium tuberculosis* under aerobic and microaerophilic conditions. Am. Rev. Respir. Dis. **114**:807–811.

70. Wayne, L. G. 1977. Synchronized replication of *Mycobacterium tuberculosis*. Infect. Immun. **17**:528–530.

71. Wayne, L. G., and Lin, K. Y. 1982. Glyoxylate metabolism and adaptation of *Mycobacterium tuberculosis* to survival under anaerobic conditions. Infect. Immun. **37**:1042–1049.

72. Wayne, L. G. 1994a. Cultivation of *Mycobacterium tuberculosis* for research purposes. In *Tuberculosis: Pathogenesis, Protection, and Control*, pp. 73–84 (Bloom, B. R., Ed.). American Society for Microbiology, Washington, D.C.

73. Wayne, L. G. 1994b. Dormancy of *Mycobacterium tuberculosis* and latency of disease. Eur. J. Clin. Microbiol. Infect. Dis. **13**:908–914.

74. Wayne, L. G., and Sramek, H. A. 1994. Metronidazole is bactericidal to dormant cells of *Mycobacterium tuberculosis*. Antimicrob. Agents Chemother. **38**:2054–2058.

75. Weis, S. E., Slocum, P. C., Blais, F. X., King, B., Nunn, M., Matney, G. B., Gomez, E., and Foresman, B. H. 1994. The effect of directly observed therapy on the rates of drug resistance and relapse in tuberculosis. N. Engl. J. Med. **330**:1179–1184.
76. World Health Organization. 1994. Framework for effective tuberculosis control. WHO/TB/94.179.
77. World Health Organization. 2000. *Global Tuberculosis Control: WHO Report 2000.* World Health Organization, Geneva.
78. World Health Organization. 2002. Global tuberculosis control: Surveillance, planning, financing. WHO Report 2002 WHO/CDS/TB/2002.295.
79. Wilson, M., DeRisi, J., Kristensen, H. H., Imboden, P., Rane, S., Brown, P. O., Schoolnik, G. K. 1999. Exploring drug-induced alterations in gene expression in *Mycobacterium tuberculosis* by microarray hybridization. Proc. Natl. Acad. Sci. USA **96**:12,833–12,838.
80. Youmans, G. P. 1979a. Morphology and metabolism. In *Tuberculosis*, pp. 8–45 (Youmans, G. P., Ed.). W. B. Saunders. Philadelphia, London, Toronto.
81. Youmans, G. P. 1979b. Pathogenesis of tuberculosis. In *Tuberculosis*, pp. 322–323 (Youmans, G. P., Ed.). W. B. Saunders. Philadelphia, London, Toronto.
82. Yuan, Y., Crane, D. D., and Barry, C. E., III. 1996. Stationary phase-associated protein expression in *Mycobacterium tuberculosis*: Function of the mycobacterial α-crystallin homolog. J. Bacteriol. **178**:4484–4492.

CHAPTER 8

Gastritis and peptic ulceration

C. Stewart Goodwin

THE DISCOVERY OF THE BACTERIAL CAUSE OF GASTRITIS

Prior to the discovery of *Helicobacter pylori*, gastritis and peptic ulceration were considered not to have a bacterial origin. This changed in April 1982, when spiral bacteria that had been observed on the gastric mucosa by Warren were cultured for the first time in the Microbiology Department of Royal Perth Hospital, Western Australia, when the author (CSG) was head of the department (31). We noted that "in old cultures coccoid bodies appeared." These coccoid forms are now thought to be the dormant phase of *H. pylori*. Among 100 patients biopsied in that study, those with gastritis and a duodenal ulcer yielded a growth of *H. pylori*. Thus when Marshall and Warren studied the medical notes, they realised that gastritis and peptic ulceration might have a bacterial origin (32).

From *Campylobacter pyloridis* to *H. pylori*

The first name given by the author to these gastric spiral bacteria was *Campylobacter pyloridis* (31). However, after six years of intensive bacteriological work by the author's research team in Western Australia, it was shown that *Campylobacter pyloridis* should be in a new genus, which he named *Helicobacter*. In 1989, in conjunction with other microbiologists in Queensland, Australia and in England, he published the new genus name; the first two species were *Helicobacter pylori*, the human stomach pathogen, and *Helicobacter mustelae*, the stomach pathogen of the ferret (20).

H. pylori as a Cause of Gastritis and Peptic Ulceration

It was soon shown that *H. pylori* is the primary cause of gastritis (21) and that *H. pylori* is an essential but insufficient cause of peptic ulceration (19). It became apparent that gastritis, found in more than 50% of the world's population, is an inflammation due to *H. pylori* in nearly every patient and is persistent and life-long. Other causes of gastritis are rare and self-limiting illnesses.

HOW *H. PYLORI* CAUSES PEPTIC ULCERATION

Peptic ulceration, and specifically duodenal ulceration, requires the interplay of several factors. If the gastric acid–producing cells in the corpus have matured, increased acid production can result from stress or smoking, and from *H. pylori* gastritis. This excess acid causes the normal columnar mucosa in the duodenum to be replaced by islands of gastric-type mucosa, called gastric metaplasia. Only in such metaplastic gastric mucosa in the duodenum can *H. pylori* from the stomach cause infection in the duodenum. When *H. pylori* duodenitis develops, then the presence of sufficient acid from the stomach frequently causes duodenal ulceration (19). In developed countries, this occurs commonly in infected people, when *H. pylori* gastritis is usually confined to the gastric antrum, and is associated with a normal or high production of stomach acid.

Why Duodenal Ulceration is Relatively Uncommon in the Developing World

H. pylori gastritis is extremely common especially in developing countries, where 80–90% of the population is infected. However, development of duodenal ulceration requires the presence of sufficient acid in the stomach, which is not always found in such patients. This is due to two factors. In developing countries, small babies and children become infected with *H. pylori*, leading to pangastritis and considerable destruction of the acid-producing cells in the corpus of the stomach (45). This results in low acid production and therefore absent duodenal ulceration. Secondly, in many developing countries poor nutrition results in production of less than sufficient stomach acid to result in duodenal ulceration.

Other Causes of Peptic Ulceration

Zollinger–Ellison syndrome is rare, but in this disease extremely high concentrations of stomach acid lead to duodenal ulceration without bacterial gastritis. Also, ingestion of non-steroidal anti-inflammatory drugs (NSAIDs) can lead to the development of gastric or duodenal ulcers, without bacterial gastritis. Some patients who are receiving NSAIDs also have gastritis due to *H. pylori*; but it is not certain whether these two conditions allow more frequent peptic ulceration than when *H. pylori* gastritis is absent in patients receiving NSAIDs.

Gastric Ulceration

Gastric ulceration has several causes. In some infected patients *H. pylori* gastritis can lead to gastric ulceration; but NSAIDs or carcinoma of the stomach can also cause gastric ulceration.

DORMANT PHASE OF *H. PYLORI* – "COCCOID FORMS"

Introduction

Soon after the first culture of *H. pylori*, coccoid forms were obtained by culture of *H. pylori* in unfavourable media conditions and were found to be very thick-walled (26). A few authors have suggested that coccoid forms are non-viable and cannot revert to spiral forms. However, there have been many reports that the coccoid form can revert to a viable spiral form. The significance of coccoids in the ecology of *H. pylori* is shown by the following facts:

- Coccoid forms can revert to spiral forms.
- Coccoids are found in the gastric mucosa.
- Animal models of *H. pylori* infection can be infected by coccoids.
- Treatment of *H. pylori* infection with antibiotics caused conversion to coccoid forms, seen in the gastric mucosa.
- Coccoid forms survive in the natural environment in water and milk and probably are important in the transmission of *H. pylori* between hosts.

ATTACHMENT TO THE GASTRIC MUCOSA BY *H. PYLORI*

The original infection in the stomach by *H. pylori* involves penetration by spiral forms of the normal mucus layer over the gastric mucosa. Thus any ingested coccoid forms must change to spiral forms for infection to occur. *H. pylori* has the greatest ability of any studied organism to penetrate normal gastric mucus (22). Other motile organisms such as *E. coli* and *Pseudomonas* lose their motility very rapidly when attempting to penetrate such mucus. *H. pylori* becomes attached to the gastric mucosal cells in a very intimate way by means of adherence pedestals (21). One of the many results of this inflammation is that production of mucus is disrupted, and the gastric mucosa loses much of its overlying mucus in areas where *H. pylori* has caused inflammation (21). Thus these areas of mucosa become more exposed to stomach acid.

The Two Methods of Attachment by *H. Pylori* to the Mucosa

Lingwood *et al.* in Canada (30) showed that one method of attachment occurs in the normal stomach when the microenvironment beneath the mucus is at a relatively neutral pH and *H. pylori* is able to attach closely to the mucosal cells. When the overlying mucus has been destroyed and the microenvironment on the surface of the mucosal cells is highly acid, *H. pylori* use a different method of attachment. This intimate adherence causes inflammation in the gastric mucosa with the production of polymorphonuclear neutrophils in the mucosa. *H. pylori* bacteria very rarely penetrate the mucosa. Dhin *et al.* (12) studied the interactions between *H. pylori* spiral and coccoid forms. Their work showed that the binding of extracellular matrix proteins may be an important mechanism of tissue adhesion, and the coccoid form of *H. pylori* can be considered an infective form in the pathogenesis of *H. pylori* infection. Coccoids bind to different carbohydrate receptors than spirals do (28).

THE NATURAL HISTORY OF *H. PYLORI* INFECTION

Nearly all people that are infected with *H. pylori* remain infected for life. The normal immune surveillance mechanisms of the body seem unable to remove this infection from the stomach mucosa. There are many reasons for this failure, but only three will be mentioned here. The bacteria initiate a neutrophil response in the mucosa, but the neutrophils must migrate to the surface of the mucosa to attack *H. pylori* because the bacteria remain on the surface of the gastric mucosa. Stomach acid destroys these neutrophils when they reach the surface of the mucosa, so they become non-viable and have no effect upon the overlying bacteria. Second, in most bacterial infections the bacterial surface is coated with the appropriate immunoglobulin, and this facilitates absorption by the immune surveillance cells of the body. However, in the acid environment of the stomach the overlying immunoglobulin becomes denatured, and *H. pylori* escapes attack. Third, when coccoid forms develop they are much less easily removed by the immune system than vegetative bacteria.

Method of Transmission of *H. pylori*

The normal method of transmission of *H. pylori* between human beings is by the faecal–oral route, and *H. pylori* can be cultured from faeces (27). Transmission seems to occur primarily in infancy and childhood, probably due to two factors. *H. pylori* bacteria remain more viable as they pass through the gut of small children than in the gut of adults and are more easily cultured from children's faeces. Secondly, children's stomach acid may be impaired by many intercurrent infections, thus allowing easier attachment of *H. pylori* to the stomach and mucosa. In some societies the mouth to mouth route may also allow transmission of *H. pylori*. It is probable that spiral and coccoid forms are transmitted.

COCCOID FORMS ARE PRESENT IN THE GASTRIC MUCOSA

Coccoid forms of *H. pylori* have been detected in the gastric mucosa by many methods. Sciortino (41) used a monoclonal antibody that reacted with the powerful urease enzyme of *H. pylori* on gastric biopsy specimens from patients with *H. pylori* gastritis. The stains showed spiral and coccoid forms within the gastric mucosa. Coccoid forms were found in stomach tissue in Sweden in nine of nine antral specimens, and five of six corpus specimens (5), in 83% of 53 Chinese patients (10) and in 11 of 14 Japanese patients with gastric cancer (39). Chan *et al.* (10) also found that the number of coccoid forms in adenocarcinoma was significantly greater than in benign peptic ulcers. In Greece, *H. pylori* was found in antral biopsy samples as coccoids and spirals (18). Coccoid bacteria are relatively common in the duodenal bulb (38). Janas *et al.* (25) found coccoid forms were present only above strongly damaged epithelial cells. Coccoids can bind to the gastric mucosa (42).

Western blots of sera from colonised patients showed that some high molecular mass antigenic fractions were expressed only in patients with coccoid forms, which suggested that coccoid forms were not a degenerative transformation and that antigens specific to the coccoid forms are expressed *in vivo* (2). An interesting study from Australia showed that strains of *H. pylori* that do not produce the characteristic urease enzyme can cause infection. Ren *et al.* (40) detected nine people who were negative by the histological biopsy test for *H. pylori* urease (the CLO-test) but were positive by PCR for the 26-kDa protein-encoding gene. Histology in these subjects showed the presence of coccoid forms in four patients, and mixed coccoid and spiral forms in three patients. Thus the gastritis produced in these patients was associated with a non-urease–producing form of *H. pylori*, and also a reduction in both local and systemic antibody levels. In another study, in biopsy specimens taken from the stomach antrum there were spiral and coccoid forms, and in five of six specimens from the corpus there were coccoid forms (5). In gastroenterological practice, endoscopy specimens are usually taken only from the antrum, and therefore coccoid forms would often be missed unless corpus specimens are taken as well.

IN VITRO STUDIES OF COCCOID FORMS

Culture Conditions Leading to *H. pylori* Coccoid Forms

Benaissa *et al.* (2) cultured *H. pylori in vitro* and noted conversion from a bacillary to a full coccoid form, via an intermediate U-shaped form. Their work suggests that the coccoid conversion is not merely a degenerative transformation and that antigens specific to the coccoid forms are expressed *in vitro*. Catrenich and Makin (6) showed that incubation for 5 days under microaerobic conditions such as are found in the stomach resulted in conversion of bacillary to coccoid forms. Oxygen concentration and pH are significant factors that induce coccoid change (49). Coccoid forms still released nitric oxide (NO) from synthetic NO generators (11). An increase in NO synthase activity observed during gastritis is thus due to the conversion of bacillary to coccoid forms, and this also shows such forms to be potentially viable.

The Morphological and Other Changes That Occur When Spiral Forms Become Coccoid Forms

When *H. pylori* changes from a bacillary to a full coccoid form there is often an intermediate U-shaped form. Organisms with a full coccoid form have a double membrane system, a polar membrane and invagination structures (2). Other culture studies have delineated the changes that occur when bacillary forms convert to the coccoid phenotype (7, 13). Extensive non-random fragmentation of ribosomal RNA occurs during conversion of *H. pylori* to the coccoid form (35).

Cell Culture

The attachment of *H. pylori* and its coccoid form to the gastric mucosa has been studied in human antral primary epithelial cells, a cell model that more closely resembles the human stomach than transformed cell lines. *H. pylori* exhibited various shapes during colonisation of these cells including spiral u-shaped, doughnut and coccoid forms (24). Coccoid forms adhere to gastric carcinoma cells (Kato III) by the same mechanisms as helical forms – adhesion pedestals, cup-like indentations and abutting adherence (52).

COCCOID FORMS IN ANIMAL MODELS

In the Gastric Mucosa

Coccoid forms have been detected in the gastric mucosa of animal models by many workers, such as in the Mongolian gerbil, where coccoid forms were found in the gastric mucosa and appeared soon after infection (48).

Coccoid Forms Can Infect Mice

Cellini *et al.* (9) were probably the first to demonstrate that coccoid forms of *H. pylori* could infect mice, and these forms were recoverable from mouse stomachs after 2 weeks of inoculation. Aleljung *et al.* (1) fed mice with spiral or coccoid forms (12 days old) of *H. pylori* strain NCTC11637 and monitored the colonisation process for 34 days post-infection. Interestingly, the coccoid form of *H. pylori* gave higher EIA absorbance values and more efficient colonisation in the mice than the spiral forms. Immunocompetent mice fed with only coccoid forms of *H. pylori* exhibited an acute inflammation process in the stomach. Wang *et al.* (53) infected mice with coccoid forms of *H. pylori* with resulting inflammation in the stomach and a significant increase in antibody response in an ELISA immunoblot after 16 weeks.

COCCOID FORMS ARE VIABLE AND CAN CHANGE TO SPIRAL FORMS

Some authors continue to doubt the relevance of coccoid forms because they maintain that they are non-viable and cannot transform to spiral forms (29, 36). However several studies have shown that coccoid forms can transform to spiral forms (4, 8). Mizoguchi *et al.* (34) studied the viability of coccoid forms of *H. pylori* and whether they are a stationary phase or a degenerative phase that leads inevitably to cell death. They found that coccoid forms conserved the capability to produce proteins for at least a 100 days when stored at 4° C in either phosphate-buffered saline or distilled water. After 20 days of storage in distilled water, exposure of the coccoids to acidic pH induced expression of several proteins that were inhibited after 20 days of storage in phosphate-buffered saline. They also determined that coccoids are more resistant to the acid environment in the stomach than spiral forms. Brenciaglia *et al.* (4) showed that coccoid forms could survive after exposure to antibiotics. Spiral forms were exposed to amoxycillin, erythromycin, gentamicin and metronidazole in culture. Coccoid forms were found after 1 to 4 weeks of culture, and the 4-week cultures of coccoid forms were cultivable into spiral forms. Sisto *et al.* (46) demonstrated the integrity of DNA and active protein synthesis in coccoid forms by analysis of the enzyme urease A and *cagA* and *vacA* genes after prolonged incubation in liquid medium. Although coccoid forms had decreased DNA and RNA concentrations after 31 days, they were not degraded and still expressed these genes. They concluded that coccoid forms are viable and therefore can act as a transmissible agent that plays a crucial role in disease relapses after antibiotic therapy. She *et al.* (43) converted three strains in culture, by exposure to metronidazole, from spiral to coccoid forms that were tested for urease activity, adherence to Hep-2 cells and Hela cells. In coccoid forms urease activity and adherence to the Hep cells and Hela cells were reduced, but the invasion of coccoid forms into the Hep-2 cells could be seen by electron microscopy. Proteins with MW>74,000 in coccoid forms decreased, but no deletion existed in amplification fragments from various genes. All this indicated that coccoid forms have potential pathogenicity. Shirai *et al.* (44) found that coccoid forms formed rapidly under anaerobic conditions, which may exist in some parts of the stomach, but acridine orange staining indicated that these forms were still viable. The cells also retained low but significant levels of the major sigma factor RpoD and also contained polyphosphate granules. All these findings indicated that coccoid forms are viable.

CHANGE TO THE COCCOID FORM DURING ANTIBIOTIC TREATMENT *IN VITRO* AND *IN VIVO*

It is highly significant that coccoid forms have been frequently found during exposure both *in vitro* and *in vivo* to antibiotics, because these coccoid forms are much more resistant to antibiotics. Thus the success of treatment may be jeopardised by this conversion to coccoid forms, allowing the organism to regrow after the end of antibiotic treatment.

In vitro studies of the conversion from spiral to coccoid forms have been done in the presence of bismuth, amoxycillin and erythromycin (3). Clarithromycin, metronidazole and amoxycillin also caused conversion to coccoid forms (4, 43, 47).

Treatment of *H. pylori* Infection and Coccoid Forms

H. pylori infection is treated with a 7–14 day course of two or more antibiotics simultaneously. However, a small proportion of patients retain the organism even after extensive and repeated multiple antibiotic treatment. In many of these patients, coccoid forms of *H. pylori* have been found that are resistant to standard antimicrobial agents (54). Treatment with clarithromycin resulted in the detection by whole cell hybridisation of coccoid forms of *H. pylori* (50). Hawkey (23) emphasised that to ensure successful elimination of *H. pylori* from patients, approaches needed to include targeting the coccoid form.

TRANSMISSION OF H. PYLORI

Detection of Coccoid Forms in Faeces

In a simulation test, faeces were inoculated with serial dilutions of either the spiral or the coccoid form of *H. pylori*. By an optimised IMS–PCR method coccoids could be detected in addition to spiral forms (37). A discrepancy between *H. pylori* stool antigen assay and urea breath test in the detection of *H. pylori* infection was found to be due to the formation of coccoid forms that appear in the faeces when bacillary forms do not appear (33). These workers developed a new enzyme immunoassay to detect *H. pylori* in stool specimens. Among 125 patients the discrepancy between the urea breath test and the stool antigen test was due to coccoid forms being detected by the immunoassay but not by the urease-based urea breath test.

Survival of Coccoid Forms in Water, Milk and Disinfectants

Under unfavourable conditions the formation of coccoid forms allows *H. pylori* to survive in the environment. *In vitro* and *in vivo* studies of coccoid forms have shown that they are a survival phenotype of *H. pylori* especially in water (51). In well water, *H. pylori* changes from the spiral form to the coccoid form after a few days. *H. pylori* can survive for up to 10 days in milk at 4° C storage, but only for 4 days in (chlorinated) tap water (16).

PCR methods have shown that coccoid forms with a different antigenicity from bacillary forms are more difficult to detect in faeces and water. Enroth and Engstrand (15) added coccoid forms to faeces and water samples, and by their modified PCR test detected the coccoid forms; and they also detected such forms in a patient's stool specimen. Engstrand (14) warned that there may be cross-reactivity between *H. pylori* and other bacteria when special PCR methods are used to detect coccoid forms in water, and such cross-reactivity may confound the DNA-based assays.

A study of disinfectant activity against different morphological forms of *H. pylori* showed that coccoid forms were viable and could withstand the action of certain disinfectants that killed helical forms (17).

CONCLUSION

The dormant phase of *H. pylori* – coccoid forms – is highly significant in the ecology of *H. pylori* infection and transmission. Coccoids can revert to helical forms and can infect animals. Coccoids are found on the gastric mucosa and may be a factor in the failure of antibiotic treatment. Coccoids can adhere to cell lines in the same way as helical forms. Coccoids survive in water, milk and disinfectants when helical forms will die and thus are significant in the transmission of *H. pylori* infection.

REFERENCES

1. Aleljung, P., Nilsson, H.-O., Wang, X., Nyberg, P., Morner, T., Warsame, I., and Wadstrom, T. 1996. Gastrointestinal colonisation of BALB/cA mice by *Helicobacter pylori* monitored by heparin magnetic separation. FEMS Immunol. Med. Microbiol. **13**:303–309.
2. Benaissa, M., Babin, P., Quellard, N., Pezennec, L., Cenatiempo, Y., and Fauchere, J. L. 1996. Changes in *Helicobacter pylori* ultrastructure and antigens during conversion from the bacillary to the coccoid form. Infect. Immun. **64**:2331–2335.
3. Bode, G., Mauch, F., and Malfertheiner, P. 1993. The coccoid forms of *Helicobacter pylori*. Criteria for their viability. Epidemiol. Infect. **111**:483–490.
4. Brenciaglia, M. I., Fornara, A. M., Scaltrito, M. M., and Dubini, F. 2000. *Helicobacter pylori*: Cultivability and antibiotic susceptibility of coccoid forms. Int. J. Antimicrob. Agents **13**:237–241.
5. Cao, J., Li, Z. Q., Borch, K., Petersson, F., and Mardh, S. 1997. Detection of spiral and coccoid forms of *Helicobacter pylori* using a murine monoclonal antibody. Clin. Chim. Acta **267**:183–196.
6. Catrenich, C. E., and Makin, K. M. 1991. Characterization of the morphologic conversion of *Helicobacter pylori* from bacillary to coccoid forms. Scand. J. Gastroenterol. Supp. **26**:58–64.
7. Cellini, L., Robuffo, I., Maraldi, N. M., and Donelli, G. 2001. Searching the point of no return in *Helicobacter pylori* life: Necrosis and/or programmed death? J. Appl. Microbiol. **90**:727–732.

8. Cellini, L., Robuffo, I., Di Campli, E., Di Bartolomeo, S., Taraborelli, T., and Dainelli, B. 1998. Recovery of *Helicobacter pylori* ATCC43504 from a viable but not culturable state: Regrowth or resuscitation? APMIS **106**:571–579.

9. Cellini, L., Allocati, N., Angelucci, D., Iezzi, T., Di Campli, E., Marzio, L., and Dainelli, B. 1994. Coccoid *Helicobacter pylori* not culturable in vitro reverts in mice. Microbiol. Immunol. **38**:843–850.

10. Chan, W.-Y., Hui, P.-K., Leung, K.-M., Chow, J., Kwok, F., and Ng, C.-S. 1994. Coccoid forms of *Helicobacter pylori* in the human stomach. Am. J. Clin. Path. **102**:503–507.

11. Cole, S. P., Kharitonov, V. F., and Guiney, D. G. 1999. Effect of nitric oxide on *Helicobacter pylori* morphology. J. Infect. Dis. **180**:1713–1717.

12. Dhin, M. M., Ringner, M., Aleljung, P., Wadstrom, T., and Ho, B. 1996. Binding of human plasminogen and lactoferrin by *Helicobacter pylori* coccoid forms. J. Med. Microbiol. **45**:433–439.

13. Donelli, G., Matarrese, P., Fiorentini, C., Dainelli, B., Taraborelli, T., Di Campli, E., Di Bartolomeo, S., and Cellini, L. 1998. The effect of oxygen on the growth and cell morphology of *Helicobacter pylori*. FEMS Microbiol. Lett. **168**:9–15.

14. Engstrand, L. 2001. *Helicobacter* in water and waterborne routes of transmission. J. Appl. Microbiol. Sympo. Supp. **90**:80S–84S.

15. Enroth, H., and Engstrand, L. 1995. Immunomagnetic separation and PCR for detection of *Helicobacter pylori* in water and stool specimens. J. Clin. Microbiol. **33**:2162–2165.

16. Fan, X.-G., Chua, A., Li, T.-G., and Zeng, Q.-S. 1998. Survival of *Helicobacter pylori* in milk and tap water. J. Gastroenterol. Hepatol. **13**:1096–1098.

17. Gebel, J., Vacata, V., Sigler, K., Pietsch, H., Rechenburg, A., Exner, M., and Kistemann, T. 2001. Disinfectant activity against different morphological forms of *Helicobacter pylori*: First results. J. Hospital Infect. **48**:S58–S63.

18. Giannios, J. N., and Karagiannis, J. A. 1996. Interactions of *Helicobacter pylori* with the gastric biological milieu. An electron microscopy study. Hellenic J. Gastroenterol. **9**:225–230.

19. Goodwin, C. S. 1988. Duodenal ulcer, Campylobacter pylori, and the "leaking roof" concept. Lancet **2**:1467–1469.

20. Goodwin, C. S., Armstrong, J. A., Chilvers, T., Peters, M., Collins, M. D., Sly, L., McConnell, W., and Harper, W. E. S. 1989. Transfer of *Campylobacter pylori* and *Campylobacter mustelae* to *Helicobacter* gen. nov. as *Helicobacter pylori* comb. nov. and *Helicobacter mustelae* comb. nov., respectively. Int. J. Systematic Bacteriol. **39**:397–405.

21. Goodwin, C. S., Armstrong, J. A., and Marshall, B. J. 1986. *Campylobacter pyloridis*, gastritis, and peptic ulceration. J. Clin. Path. **39**:353–365.

22. Hazell, S. L., Lee, A., Brady, L., and Hennessy, W. 1986. *Campylobacter pyloridis* and gastritis: Association with intercellular spaces and adaptation to an environment of mucus as important factors in colonization of the gastric epithelium. J. Infect. Dis. **153**:658–663.

23. Hawkey, C. J. 1997. Treatment of *Helicobacter pylori*. Emerging Drugs **2**:305–325.

24. Heczko, U., Smith, V. C., Meloche, R. M., Buchan, A. M. J., and Finlay, B. B. 2000. Characteristics of *Helicobacter pylori* attachment to human primary antral epithelial cells. Microbes Infect. **2**:1669–1676.

25. Janas, B., Czkwianianc, E., Bak-Romaniszyn, L., Bartel, H., Tosik, D., Planeta-Malecka, I. 1995. Electron microscopic study of association between coccoid forms of *Helicobacter pylori* and gastric epithelial cells. Am. J. Gastroenterol. **90**:1829–1833.

26. Jones, D. M., Curry, A., and Fox, A. J. 1985. An ultrastructural study of the gastric campylobacter-like organism '*Campylobacter pyloridis*.' J. Gen. Microbiol. **131**:2335–2341.

27. Kelly, S. M., Pitcher, M. C., Farmery, S. M., and Gibson, G. R. 1994. Isolation of *Helicobacter pylori* from feces of patients with dyspepsia in the United Kingdom. Gastroenterology **107**:1671–1674.

28. Khin, M. M., Hua, J. S., Ng, H. C., Wadstrom, T., and Ho, B. 2000. Agglutination of *Helicobacter pylori* coccoids by lectins. World J. Gastroenterol. **6**:202–209.

29. Kusters, J. G., Gerrits, M. M., Van Strijp, J. A. G., and Grauls, C. M. J. E. 1997. Coccoid forms of *Helicobacter pylori* are the morphologic manifestation of cell death. Infect. Immun. **65**:3672–3679.

30. Lingwood, C. A., Wasfy, G., Han, H., and Huesca, M. 1993. Receptor affinity purification of a lipid-binding adhesin from *Helicobacter pylori*. Infect. Immun. **61**:2474–2478.

31. Marshall, B. J., Royce, H., Annear, D. I., Goodwin, C. S., Pearman, J. A., Warren, J. R., and Armstrong, J. A. 1984. Original isolation of *Campylobacter pyloridis* from human gastric mucosa. Microbios Lett. **25**:83–88.

32. Marshall, B. J., and Warren, J. R. 1984. Unidentified curved bacilli in the stomach of patients with gastritis and peptic ulceration. Lancet **1**:1311–1315.

33. Masoero, G., Lombardo, L., Della, M. P., Vicari, S., Crocilla, C., Duglio, A., and Pera, A. 2000. Discrepancy between *Helicobacter pylori* stool antigen assay and urea breath test in the detection of *Helicobacter pylori* infection. Digest. Liver Dis. **32**:285–290.

34. Mizoguchi, H., Fujioka, T., and Nasu, M. 1999. Evidence for viability of coccoid forms of *Helicobacter pylori*. J. Gastroenterol. **34**:32–36.

35. Monstein, H.-J., Tiveljung, A., and Jonasson, J. 1998. Non-random fragmentation of ribosomal RNA in *Helicobacter pylori* during conversion to the coccoid form. FEMS Immunol. Med. Microbiol. **22**:217–224.

36. Nabwera, H. M., and Logan, R. P. H. 1999. Epidemiology of *Helicobacter pylori:* Transmission, translocation and extragastric reservoirs. J. Physiol. Pharmacol. **50**:711–722.

37. Nilsson, I., Utt, M., Nilsson, H.-O., Ljungh, A., and Wadstrom, T. 2000. Two-dimensional electrophoretic and immunoblot analysis of cell surface proteins of spiral-shaped coccoid forms of *Helicobacter pylori*. Electrophoresis 21:2670–2677.

38. Noach, L. A., Rolf, T. M., and Tytgat, G. N. J. 1994. Electron microscopic study of association between *Helicobacter pylori* and gastric and duodenal mucosa. J. Clin. Path. 47:699–704.

39. Ogata, M., Araki, K., and Ogata, T. 1998. An electron microscopic study of *Helicobacter pylori* in the surface mucous gel layer. Histolo. Histopath. 13:347–358.

40. Ren, Z., Musicka, M., Pang, G., Dunkley, M., Batey, R., Beagley, K., Routley, D., Clancy, R., and Russell, A. 2000. Non-urease producing *Helicobacter pylori* in chronic gastritis. Australian & New Zealand J. Med. **30**:578–584.

41. Sciortino C. V., Jr. 1993. An immunofluorescent stain for *Helicobacter pylori*. Hybridoma 12:333–342.

42. Segal, E. D., Falkow, S., and Tompkins, L. S. 1996. *Helicobacter pylori* attachment to gastric cells induces cytoskeletal rearrangements and tyrosine phosphorylation of host cell proteins. Proc. Natl. Acad. Sci. USA **93**:1259–1264.

43. She, F. F., Su, D. H., Lin, J. Y., and Zhou, L. Y. 2001. Virulence and potential pathogenicity of coccoid *Helicobacter pylori* induced by antibiotics. World J. Gastroenterol. **7**:254–258.

44. Shirai, M., Kakada, J., Shibata, K., Morshed, M. G., Matsushita, T., and Nakazawa, T. 2000. Accumulation of polyphosphate granules in *Helicobacter pylori* cells under anaerobic conditions. J. Med. Microbiol. **49**:513–519.

45. Sipponen, P., Kekki, M., and Siurala, M. 1991. The Sydney System: epidemiology and natural history of chronic gastritis. J. Gastroenterol. Hepatol. **6**:244–251.

46. Sisto, F., Brenciaglia, M. I., Scaltrito, M. M., and Dubini, F. 2000. *Helicobacter pylori:* ureA, cagA and vacA expression during conversion to the coccoid form. Int. J. Antimicrob. Agents **15**:277–282.

47. Sorberg, M., Nilsson, M., Hanberger, H., and Nilsson, L. E. 1996. Morphologic conversion of *Helicobacter pylori* from bacillary to coccoid form. Eur. J. Clin. Microbiol. Infectious Diseases **15**:216–219.

48. Sugiyama, A., Ishida, K., Ikeno, T., Maruta, F., Kawasaki, S., Akamatsu, T., and Katsuyama, T. 1998. Distinction of the shape of *Helicobacter pylori* using stereo pairs constructed from digitized light microscopic images. Digest. Dis. Sci. **43**:188S–191S.

49. Tominaga, K., Hamasaki, N., Watanabe, T., Uchida, T., Fujiwara, Y., Takaishi, O., Higuchi, K., Arakawa, T., Ishii, E., Kobayashi, K., Yano, I., and Kuroki, T. 1999. Effect of culture conditions on morphological changes of *Helicobacter pylori*. J. Gastroenterol. **34**:28–31.

50. Trebesius, K., Panthel, K., Strobel, S., Vogt, K., Faller, G., Kirchner, T., Kist, M., Heesemann, J., and Haas, R. 2000. Rapid and specific detection of *Helicobacter pylori* macrolide resistance in gastric tissue by fluorescent in situ hybridisation. Gut **46**:608–614.

51. Velazquez, M., and Feirtag, J. M. 1999. *Helicobacter pylori*: Characteristics, pathogenicity, detection methods and mode of transmission implicating foods and water. Int. J. Food Microbiol. **53**:95–104.

52. Vijayakumari, S., Khin, M. M., Jiang, B., and Ho, B. 1995. The pathogenic role of the coccoid form of *Helicobacter pylori*. Cytobios **82**:251–260.

53. Wang, X., Sturegard, E., Rupar, R., Nilsson, H.-O., Aleljung, P. A., Carlen, B., Willen, R., and Wadstrom, T. 1997. Infection of BALB/c A mice by spiral and coccoid forms of *Helicobacter pylori*. J. Med. Microbiol. **46**:657–663.

54. Wen, M., Yamada, N., Zhang, Y., and Matsuhisa, T. 1997. Morphological changes of *Helicobacter pylori* after antibacterial therapy: An electron microscope study. Med. Electron Microscopy **30**:131–137.

CHAPTER 9

Resumption of yeast cell proliferation from stationary phase

Gerald C. Johnston and Richard A. Singer

Populations of non-proliferating cells are often referred to as being in a quiescent, or G0, state. For unicellular microorganisms this quiescent state is a regulated response to a limitation of the nutrient supply, whereas for multicellular organisms, quiescence can also be induced by hormones, growth factors or contact with other cells. Indeed, most cells, including those in multicellular organisms, are found in a quiescent state with respect to cell proliferation. Nevertheless, we understand relatively little about the quiescent state of eukaryotic cells. Genetic and molecular approaches with the unicellular budding yeast *Saccharomyces cerevisiae* offer an opportunity to investigate this important aspect of eukaryotic biology.

The budding yeast *S. cerevisiae* is widely used to investigate fundamental cellular activities, for two important reasons. First is the genetic and molecular facility that the yeast system provides. For example, classical (Mendelian) genetic analysis and mutant isolation are straightforward with this organism, and many strategies are available for gene cloning. Moreover, yeast offers the ability to easily and routinely transfer cloned genes back into the genome to replace the resident chromosomal gene in the proper chromosomal context, an approach that allows elucidation of gene-product function in an *in vivo* context (reviewed in ref. 9). Furthermore, the fact that the entire yeast genome, and thus the complement of yeast proteins, is known (8) provides a comprehensive perspective for functional studies. The yeast system has therefore been invaluable for the application of molecular approaches to the study of eukaryotic gene function. Second, the yeast system is highly relevant to broader concerns: from yeast to mammalian cells there is demonstrated *functional* conservation (3, 13; see 4) of the components of many fundamental processes, including the regulation of cell proliferation, growth, gene expression, signal transduction and intracellular membrane trafficking through vesicular transport.

STATIONARY PHASE

Two major developmental transitions for cells are the shifts between the state of active cell proliferation and a non-proliferating or quiescent condition. For yeast, a depleted nutrient supply causes proliferating cells to embark on a developmental process that will ultimately lead to a non-proliferating state often referred to as stationary phase (Fig. 9.1). The response of actively proliferating yeast cells to nutrient depletion has perhaps been best characterized for limitation of carbon supply (typically glucose). As cells become limited for glucose, the rate of cell proliferation slows and eventually ceases, and cells develop thickened cell walls, become optically refractile and undergo changes in both physiology and gene expression (reviewed in ref. 26). As is typical for most non-proliferating cells, stationary-phase yeast cells are arrested with an unreplicated complement of DNA, with cells exhibiting a characteristic morphology referred to as unbudded. (For this yeast the process of cell division is initiated by the production of a bud, the incipient daughter cell, at a time that approximates the onset of DNA replication. Cells that become arrested with an unreplicated complement of DNA do not exhibit a bud and are termed unbudded.)

As cells become limited for nutrients, metabolic processes slow and there is a rapid and regulated decline in both transcription and translation, with a major shutdown in the transcription of ribosomal protein genes. Nevertheless, the mRNAs for some proteins such as some of the heat-shock genes (i.e., *UBI4*) actually increase. Likewise, levels of the storage carbohydrates glycogen and trehalose accumulate dramatically and lipid vesicles accumulate in the cytoplasm. In addition, yeast cells become markedly resistant to high temperatures. Such stationary-phase cells also display a remarkable ability to survive for long periods (at least months) without added nutrients (26).

With the restoration of nutrients, starved stationary-phase cells undergo a developmental transition in preparation for the onset of cell proliferation that includes a loss of many of these stationary-phase properties. Nevertheless, resumption of cell proliferation from stationary phase is not a simple reversal of processes that led to stationary phase in the first place. Unique requirements in yeast cells for the developmental transition from quiescence to active cell proliferation have been revealed by genetic and molecular approaches, as described here.

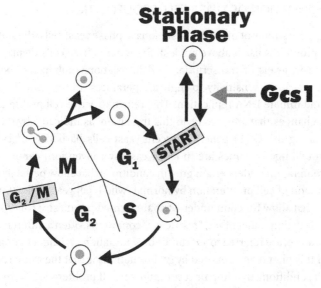

Figure 9.1. Active cell proliferation and the differentiated state termed stationary phase. Cell proliferation is carried out by performance of the cell cycle, which is marked by the periodic replication of nuclear DNA during S phase, followed by the segregation of replicated DNA by mitosis and cytokinesis, during M phase. These two phases are separated by the G_1 and G_2 phases, in which checkpoint regulation ensures that the cell is properly prepared for replication and for mitosis; these checkpoints are termed Start and G_2/M, respectively (see ref. 12). For the budding yeast *Saccharomyces cerevisiae* discussed here, the major regulatory point is Start (11). Also shown is the yeast morphological cycle. A yeast cell of this type reproduces by the production of a bud that increases in size as the cell progresses through the cell cycle. One useful feature of this budding behaviour is that bud morphology reflects cell-cycle position, so that a cell without a bud (an unbudded cell) is in the prereplicative G_1 phase, initiation of a bud roughly coincides with the onset of S phase and bud size continues to enlarge as a cell progresses through the cell cycle so that at the onset of mitosis the bud is approaching the size of the parent cell. Yeast cell growth and cell-cycle activity is responsive to the nutrient supply. Nutrient depletion does not prevent a cell from completing the cell cycle but does affect the initiation of a new cell cycle at Start. A nutrient-limited cell may then cease cell proliferation and undergo the developmental transition to stationary phase. Upon nutrient replenishment, a stationary-phase cell returns to the cell cycle, and it is this developmental transition, the return to active cell proliferation, that can be specifically impaired by the absence of Gcs1 protein function.

RESUMPTION OF YEAST CELL PROLIFERATION

Upon resupply of nutrients to stationary-phase yeast cells, the stationary-phase properties listed above are lost. Stationary-phase cells display a rapid increase in levels of transcription and translation, mobilize carbohydrate stores, reacquire sensitivity to high temperatures and eventually produce buds and initiate DNA replication. This return to active cell proliferation requires changes that are specific to this transition, as indicated by the effects of mutating the *GCS1* gene. Quiescent yeast cells harboring a *gcs1* mutation (mutant-gene names are in lower case by convention) cannot resume cell proliferation under certain growth conditions that allow perfectly normal resumption of cell proliferation by normal (wild-type) yeast cells. (Such mutations that allow function under what are termed permissive conditions but impair function under other, "restrictive" conditions, often a different growth temperature, are referred to as "conditional mutations.") The remarkable feature of this phenotype imposed by *gcs1* mutations is that the same restrictive growth conditions that impair resumption of cell proliferation from stationary phase have *no* inhibitory effect on the ongoing proliferation of *gcs1* mutant cells that have been allowed to resume active cell proliferation under permissive conditions; they also have no detectable effect on the successful transition *into* stationary phase by *gcs1* cells (6, 7, 16). Thus the effects of a *gcs1* mutation, even under restrictive conditions, are evident only during the transition to active cell proliferation (Fig. 9.1). These *gcs1* effects are manifested at the low growth temperature of 15° C (yeast cells grow well throughout a wide range of temperatures, making temperature-conditional mutations such as *gcs1* useful for elucidating important cellular processes) and can be easily brought about (or alleviated) by simple temperature shifts. The *gcs1* mutations are therefore important tools in understanding the special requirements for the transition from quiescence to active cell proliferation.

The conditional (cold-sensitive) blockage of the resumption of cell proliferation for *gcs1* mutant cells has been characterized to some extent. These findings indicate that the impediment to the resumption of cell proliferation that is manifested by a *gcs1* mutation arises *early* in the development of stationary-phase properties, for *gcs1* mutant cells become unable to resume proliferation soon after nutrient deprivation takes hold (6, 7, 14). On the other hand, this degree of deprivation is sufficient to trigger the accumulation of storage carbohydrates and the acquisition of thermo-tolerance, two hallmarks of stationary-phase development (for reviews, see refs. 16, 26). What happens upon nutrient resupply to starved, and consequently cold-sensitive, *gcs1* mutant cells has also been investigated. At the restrictive 15° C temperature, nutrient resupply to starved *gcs1* mutant cells stimulates transcription of at least some genes, degradation of storage carbohydrates and loss of the stationary-phase property of thermo-tolerance; the kinetics of these responses are similar to those seen for normal cells. However, the *gcs1* mutant cells fail to re-enter the mitotic cell cycle as indicated by morphological and genetic criteria (6, 7, 14). This behavior suggests that the *gcs1* mutation does not affect the ability of a starved cell to sense and respond to nutrients and to initiate the de-differentiation from the stationary-phase state but does impair a process that is needed for subsequent cellular development.

The intracellular process that is directly affected by a *gcs1* mutation is suggested by the characteristics of the protein product of the normal *GCS1* gene. We have characterized the *GCS1* gene at the molecular level (14) and demonstrated that Gcs1 (the *GCS1* gene product) is a GTPase-activating protein (GAP) for the Arf type of monomeric GTP-binding proteins (18). For eukaryotic cells, Arf proteins regulate various stages of the core intracellular process of vesicular transport (reviewed in refs. 2, 5), and our findings indicate that the Gcs1 protein and its mammalian functional homolog are also central components of vesicular transport (18, 19).

VESICULAR TRANSPORT AND THE GCS1 PROTEIN

Vesicular transport is a multi-stage process through which integral membrane proteins and other components are moved, within a cell, among several membrane compartments, including the plasma membrane and the endosome/lysosome complex (reviewed in refs. 21, 22; Fig. 9.2). The mechanism of this transport, at each stage, is considered to involve common processes of vesicle formation and fusion. Initiating this overall process, a localized region of the membrane of an intracellular compartment rounds up and buds off to form a membrane-bound transport vesicle; this vesicle then fuses with the membrane of a "target" compartment. This vesicular transport is mediated in part by Arf proteins, which in the GTP-bound form stimulate the recruitment of "coat proteins" to a membrane for vesicle budding (23). The subsequent removal of these coat proteins may also be mediated by Arf proteins, through the hydrolysis of Arf-bound GTP to GDP. The regulated hydrolysis of Arf-bound GTP may also be involved in the packaging of "cargo" proteins into the nascent vesicle (10). However, Arf proteins, although members of the *ras* superfamily of small G-proteins and closely related to these GTPase enzymes in structure, have no intrinsic GTPase activity (17), so that GTP hydrolysis by Arf protein needs the participation of a GAP such as Gcs1. Gcs1 was the first yeast ArfGAP identified, and mammalian cells contain

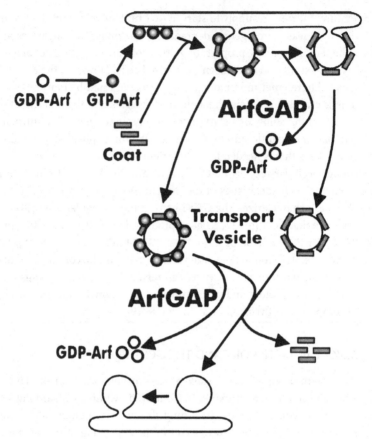

Figure 9.2. Vesicular transport in brief. The transport of material between membrane compartments, including the plasma membrane, takes place by the formation of a transport vesicle at a donor compartment followed by the fusion of the vesicle with a target membrane. Vesicle formation involves the recruitment of several protein components, including coat proteins, to a membrane to produce a so-called "coated vesicle" carrying appropriate cargo material (proteins and membrane lipids). The coat is then removed before fusion of the transport vesicle with a target membrane. Transport-vesicle formation involves activation of a type of monomeric GTPase protein, termed Arf (open circle), and by the exchange of bound GDP for GTP, a process mediated by a guanine-nucleotide exchange factor (GEF). The GTP-bound form of Arf (filled circle) associates with the membrane and recruits additional coat proteins (rectangles), causing membrane deformation and vesicle formation. During this process proteins and lipids destined for transport become incorporated into the nascent vesicle. Both cargo packaging during vesicle formation and removal of the coat as a prelude to vesicle fusion and cargo delivery require stimulation of the GTPase activity of Arf, a function performed by a GTPase-activating protein (GAP). The Gcs1 and Age2 proteins discussed here are two such GAPs.

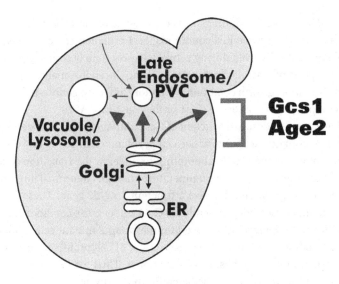

Figure 9.3. Vesicular transport in a cellular context. Within a cell there are several transport stages involving the endoplasmic reticulum (ER), the Golgi apparatus, the plasma membrane, the vacuole (the yeast equivalent of the lysosome) and the endosomal complex (simplified here). Each of these stages is mediated by the production of transport vesicles. Transport from the *trans*-Golgi network (TGN) takes several routes, including transport to the vacuole, the plasma membrane and the endosomal compartment. The Gcs1 and Age2 ArfGAP proteins modulate this transport out of the TGN.

several structural homologs of Gcs1, including one that can functionally re-place Gcs1 for yeast cell growth (19).

Yeast cells also have several structurally related Gcs1-like proteins, and we have found that at least two of these related proteins, like Gcs1, mediate vesicular transport (19). One of these proteins, Age2, has *in vivo* functions that overlap those of Gcs1. Cells missing either Gcs1 or Age2 protein can grow relatively well, whereas cells missing both of these proteins are dead (20, 27 [in which the Age2 protein is referred to as Sat2]). Our recent studies point to an overlapping essential function for Gcs1 and Age2 in endosomal traffic (Fig. 9.3).

ENDOSOMAL FUNCTION AND THE GCS1 PROTEIN

Vesicular transport through the endosomal complex regulates the re-moval and sorting of material from the plasma membrane for delivery to organelles such as the vacuole (the yeast equivalent of the lysosome) or for recycling back to the plasma membrane (reviewed in ref. 25). An endoso-mal trafficking function is shared between the Gcs1 and Age2 proteins, as indicated by our genetic findings. For cells deleted for the *AGE2* gene and thus devoid of the Age2 protein, a conditional *gcs1* mutation can cause *gcs1 age2* double-mutant cells, even during active cell growth, to become impaired for several aspects of vesicular transport involving the endosomal compartment (20). Under most growth conditions, therefore, effective endosomal function can be supported by either the Gcs1 or the Age2 ArfGAP.

The shared endosomal function of Gcs1 and Age2 does not extend to the transition from quiescence to active cell proliferation at 15° C, however. We found that *gcs1* mutant cells attempting to undergo the transition to active cell proliferation at a restrictive temperature show an endosomal impairment (24), even though these cells have a functional *AGE2* gene. These findings imply an important endosomal role for Gcs1 under these conditions. In contrast, no such impairment has been found for *age2* mutant cells under any growth conditions, including the resumption of cell proliferation from stationary phase (our unpublished observations). Thus the critical endosomal role during this transition is mediated by an activity specific to Gcs1 under these conditions.

The involvement of two ArfGAPs suggests a mechanism that leads to the *gcs1*-mediated conditional defect in resumption of cell proliferation from stationary phase. In the absence of the Gcs1 protein, Age2-mediated processes are somehow unable (at 15° C) to carry out the vesicular transport necessary for the transition from quiescence to active cell proliferation. Of course, for cells containing Gcs1 protein this inherent cold sensitivity of Age2-mediated processes during this transition is masked by the effective function of Gcs1. The loss of Gcs1 function due to a *gcs1* mutation therefore unmasks the inherently cold-sensitive Age2-mediated process, thus bringing about the *gcs1* mutational effects. The Age2 ArfGAP may not itself be the component of vesicular transport that is cold sensitive for function, simply because Age2 (in the absence of Gcs1) supplies essential function for mitotic growth at the restrictive temperature. Instead, some component that facilitates Age2 function (but not Gcs1 function) for the resumption of cell proliferation may be inherently cold sensitive. The identification and characterization of this cold-sensitive (presumably) endosomal function that affects the resumption of yeast cell proliferation will provide insight into specific activities required for this important developmental transition.

The fact that components of vesicular transport in general, and endosomal function in particular, can have a specific effect on the transition from quiescence to active cell proliferation leads us to suggest that an important activity for this transition may be intracellular remodeling. Indeed, the intracellular membrane morphology of starved, quiescent yeast cells is distinctly different (more vesiculated and "chaotic") than that of normally growing cells, and *gcs1* mutant cells, upon nutrient resupply at 15° C, are unable to remodel intracellular membranes or the actin cytoskeleton, to acquire a normal morphology (our unpublished data). This deficit in membrane remodeling may reflect several potential effects of Gcs1. Impaired Gcs1 activity can derange cytoskeletal regulation (1), while Gcs1-related mammalian proteins, and Arf proteins, similarly affect actin dynamics and may also have additional effects (reviewed in ref. 15). Alternatively, impaired Gcs1 activity may interfere with the redistribution of proteins and membrane components through vesicular-transport deficiencies. It may be that a mislocalization of certain key cargo proteins (targeted and localized to specific membrane compartments by vesicular transport), rather than a global inadequacy in vesicular transport itself, contributes to the cold sensitivity seen in the absence of Gcs1 protein. Predicting the identity of such key cargo proteins is difficult, for the spectrum of cargo translocated *via* endosomal functions is poorly defined. However, as an example, a lack of Gcs1 function during the transition from quiescence to active cell proliferation (at low growth temperatures) may prevent the necessary removal, by endocytic processes, of inhibitory proteins from the plasma membrane of the cell. Further analysis of the overlapping essential function of Gcs1 + Age2 has the potential to identify such critical proteins or other cargo (perhaps critical plasma-membrane components) and in more general terms to address whether vesicular transport itself or a related ArfGAP-mediated activity, such as cytoskeletal regulation, is critical for the transition to active cell proliferation. In any case, the overwhelming evidence from many studies tells us that these investigations with yeast will reveal activities that specifically facilitate the transition from quiescence to cell proliferation.

ACKNOWLEDGEMENTS

Thanks to Pak Poon for providing the figures. Our work described here was supported by a grant from the National Cancer Institute of Canada with funds from the Canadian Cancer Society.

REFERENCES

1. Blader, I. J., Cope, M. J. T. V., Jackson, T. R., Profit, A. A., Greenwood, A. F., Drubin, D. G., Prestwich, G. D., and Theibert, A. B. 1999. *GCS1*, an Arf guanosine triphosphatase-activating protein in *Saccharomyces cerevisiae*, is required for normal actin cytoskeletal organization in vivo and stimulates actin polymerization in vitro. Mol. Biol. Cell **10**:581–596.

2. Boman, A. L., and Kahn, R. A. 1995. Arf proteins: The membrane traffic police? Trends Biochem. Sci. **20**:147–150.

3. Botstein, D., Chervitz, S. A., and Cherry, J. M. 1997. Yeast as a model organism. Science **277**:1259–1260.

4. Chervitz, S. A., Aravind, L., Sherlock, G., Ball, C. A., Koonin, E. V., Dwight, S. S., Harris, M. A., Dolinski, K., Mohr, S., Smith, T., Weng, S., Cherry, J. M., and Botstein, D. 1998. Comparison of the complete protein sets of worm and yeast: Orthology and divergence. Science **282**:2022–2028.

5. Donaldson, J. G., and Klausner, R. D. 1994. ARF: A key regulatory switch in membrane traffic and organelle structure. Curr. Opin. Cell Biol. **6**:527–532.

6. Drebot, M. A., Johnston, G. C., and Singer, R. A. 1987. A yeast mutant conditionally defective only for reentry into the mitotic cell cycle from stationary phase. Proc. Natl. Acad. Sci. USA **84**:7948–7952.

7. Drebot, M. A., Barnes, C. A., Singer, R. A., and Johnston, G. C. 1990. Genetic assessment of stationary phase for cells of the yeast *Saccharomyces cerevisiae*. J. Bacteriol. **172**:3584–3589.

8. Dujon, B. 1996. The yeast genome project: What did we learn? Trends Genet. **12**:263–270.

9. Guthrie, C., and Fink, G. R. (Eds.) 1991. *Guide to Yeast Genetics and Molecular Biology*. Meth. Enzymol. **194**:21–131.

10. Goldberg, J. 2000. Decoding of sorting signals by coatomer through a GTPase switch in the COPI coat complex. Cell **100**:671–679.

11. Hartwell, L. H. 1974. *Saccharomyces cerevisiae* cell cycle. Bacteriol. Rev. **38**: 164–198.

12. Hartwell, L. H., and Weinert, T. A. 1989. Checkpoints: Controls that ensure the order of cell cycle events. Science **246**:629–634.

13. Hartwell, L. H., Szankasi, P., Roberts, C. J., Murray, A. W., and Friend, S. H. 1997. Integrating genetic approaches into the discovery of anticancer drugs. Science **278**:1064–1068.

14. Ireland, L. S., Johnston, G. C., Drebot, M. A., Dhillon, N., DeMaggio, A. J., Hoekstra, M. F., and Singer, R. A. 1994. A member of a novel family of yeast 'Zn-finger' proteins mediates the transition from stationary phase to cell proliferation. EMBO J. **13**:3812–3821.

15. Jackson, T. R., Kearns, B. G., and Theibert, A. B. 2000. Cytohesins and centaurins: Mediators of PI 3-kinase-regulated Arf signaling. Trends Biochem. Sci. **25**:489–495.

16. Johnston, G. C., and Singer, R. A. 1990. Regulation of proliferation by the budding yeast *Saccharomyces cerevisiae*. Biochem. Cell Biol. **68**:427–435.

17. Kahn, R. A., and Gilman, A. G. 1986. The protein cofactor necessary for ADP ribosylation of G_s by cholera toxin is itself a GTP binding protein. J. Biol. Chem. **261**:7906–7911.

18. Poon, P. P., Wang, X., Rotman, M., Huber, I., Cukierman, E., Cassel, D., Singer, R. A., and Johnston, G. C. 1996. *Saccharomyces cerevisiae* Gcs1 is an ADP-ribosylation factor GTPase-activating protein. Proc. Natl. Acad. Sci. USA **93**:10,074–10,077.

19. Poon, P. P., Cassel, D., Spang, A., Rotman, M., Pick, E., Singer, R. A., and Johnston, G. C. 1999. Retrograde transport from the yeast Golgi is mediated by two ARF GAP proteins with overlapping function. EMBO J. **18**:555–564.

20. Poon, P. P., Nothwehr, S. F., Singer, R. A., and Johnston, G. C. 2001. The Gcs1 and Age2 ArfGAP proteins provide essential overlapping function for transport from the yeast *trans*-Golgi network. J. Cell Biol. **155**:1239–1250.

21. Rothman, J. E. 1994. Mechanisms of intracellular protein transport. Nature **372**:55–63.

22. Rothman, J. E., and Wieland, F. T. 1996. Protein sorting by transport vesicles. Science **272**:227–234.

23. Schekman, R., and Orci, L. 1996. Coat proteins and vesicle budding. Science **271**:1526–1533.

24. Wang, X., Hoekstra, M. F., DeMaggio, A. J., Dhillon, N., Vancura, A., Kuret, J., Johnston, G. C., and Singer, R. A. 1996. Prenylated isoforms of yeast casein kinase I, including the novel Yck3p, suppress the *gcs1* blockage of cell proliferation from stationary phase. Mol. Cell. Biol. **16**:5375–5385.

25. Wendland, B., Emr, S. D., and Riezman, H. 1998. Protein traffic in the yeast endocytic and vacuolar protein sorting pathways. Curr. Opin. Cell Biol. **10**:513–522.

26. Werner-Washburne, M., Braun, E., Johnston, G. C., and Singer, R. A. 1993. Stationary phase in the yeast *Saccharomyces cerevisiae*. Microbiol. Rev. **37**:383–401.

27. Zhang, C.-J., Cavenaugh, M. M., and Kahn, R. A. 1998. A family of Arf effectors defined as suppressors of the loss of Arf function in the yeast *Saccharomyces cerevisiae*. J. Biol. Chem. **273**:19,792–19,796.

CHAPTER 10

Resting state in seeds of higher plants: dormancy, persistence and resilience to abiotic and biotic stresses

Hugh W. Pritchard and Peter E. Toorop

INTRODUCTION

Changes in the physical environment, particularly in relation to temperature, light, humidity and gaseous atmosphere, limit the growth of all organisms and promote entry into a long-term persistent, "dormant" state. Of these, probably moisture and temperature are the most important. At high humidities (>99%), organisms tend to grow at an unhindered rate. However, dehydration towards 90% relative humidity (RH) severely reduces metabolism and growth of plants, fungi, yeast and animals. For example, nematodes coil and *Fusarium hyphae* become stationary when RH is reduced (45). To facilitate survival of such high levels of desiccation, a complex sequence of biochemical changes is needed that safely downregulates biosynthetic pathways. For example in seeds, the accumulation of pernicious compounds (such as free radicals/activated oxygen and lipid peroxidation products) is avoided, and/or counteracted by the accumulation of protectants like tocopherols (60). In seeds, the point of growth cessation has been estimated to be around -2 to -4 MPa ($\sim$99% RH). Seeds fail to germinate below this water potential, although metabolism, e.g., respiration, only ceases at about -15 MPa ($\sim$90% RH) (73). Below this water potential, insufficient water means relative metabolic stasis and enhanced longevity as further moisture is withdrawn down to around -250 MPa ($\sim$15% RH). It is not just the seeds (sporophytes) of many higher plants which can tolerate such an environment, but also vegetative tissue (gametophyte) of a small number of plants (the resurrection plants). In addition, spores of pteridophytes (see (8)), fungi, microbes (especially bacterial endospores) and animal cysts (45) can survive high levels of desiccation. In general, then, it is the reproductive "propagules" of organisms that have the greatest tolerance of desiccation and, as a consequence, enhanced capability to become persistent. Nonetheless, some organisms persist in a dormant state at high water potentials (i.e., in hydrated tissue/cells). This is the result of a block in metabolism which hinders growth, for example the vegetative propagules and buds of higher plants (42, 87). However, greater attention has been paid to seed dormancy than other dormant plant forms because of its key ecological role in the survival of the species in a spatial and temporal context (3, 26). The main purpose of this review is to compare aspects of "dormancy" in the seeds of higher plants with the quiescent non-replicating or slowly growing state in organisms of other kingdoms, particularly microbes.

DORMANCY IN HIGHER PLANTS

Dormancy in higher plants is not restricted to their reproductive propagules but is present in diverse structures, such as axillary and apical buds and vegetative propagules (corms, bulbs, tubers). Dormancy varies in depth both temporally, with season, and spatially, between tissues/organs of the same plant. There is a tendency to associate this ability primarily with temperate environments, as the environmental cues for the deactivation and reactivation of the dormant state are well established. However, seasonal dormancy in vegetative tissues is also a feature of many tropical environments (42, 87).

The complexity of this trait has precluded the development of an unambiguous and agreed upon definition of dormancy in plants. A popular definition is "a temporary suspension of visible growth of any structure containing a meristem," i.e., a group of potentially actively dividing cells. Dormancy also includes the "dynamic" change of growth of the primordia to initiate and develop special dormant organs before the temporary suspension of growth (42, 56, 87). This latter definition of dormancy permits one growth cycle to occur per year and limits the continuity of the growth cycle genetically and environmentally (56).

Seed Dormancy Categories

There have been numerous attempts to categorise seed dormancy (3, 43, 54). Many have focussed on descriptions of the environmental blocks to germination, such as light-induced (photo-) dormancy and high temperature (thermo-) dormancy (29) or, conversely, the conditions required to facilitate dormancy loss (3). Many factors are linked with dormancy loss or germination initiation, but there is usually little indication of the underlying mechanisms.

This has made it difficult, although not impossible, to separate dormancy relief from the initiation of germinative growth (13, 30, 69).

Using a range of physiological and morphological features as descriptors of dormancy, Baskin and Baskin (1998) have been able to classify seed dormancy for more than 3000 species. This classification system consists of six types, two of which, namely the physical and mechanical types, are rather similar (Table 10.1). The presence of a hard seed coat provides a mechanical resistance to embryo growth and/or impermeability to water or oxygen which are needed for dormancy loss and germination. Well known species, such as numerous legumes, sumaks and olives, fit one or both of these categories. The other four types of dormancy are physiological, a classification which contains three levels, morphological, morpho-physiological and chemical. However, there is little convincing evidence for the chemical category since many studies have been confounded by the presence of physiological dormancy in the seeds (3).

Seeds with physiological dormancy are mostly permeable to water, but the embryo is inhibited from growing. Inhibition may be due to the covering structures, which impair oxygen permeation and sometimes block leakage of an inhibitor. This type of dormancy restricts germination to a very narrow range of temperatures, and in some species no germination takes place at all. Dormancy can be overcome, usually, by treating the seeds with relatively short periods (up to 12 weeks) of cold, longer periods of dry warmth (after-ripening) or with wet warmth (weeks to months). The depth of dormancy can vary considerably between species and between seed lots of the same species, and the effectiveness of dormancy-breaking chemicals (e.g., KNO_3 and gibberellin) on intact seeds (dispersal units) is dose-dependent. Light is usually required to stimulate germination and may interact synergistically with dormancy-breaking chemicals and alternating temperatures. Well known species that fit into this broad category of dormancy are weeds, many garden flower species and some woody plants, such as maple and apple. Physiological dormancy is relatively easily imposed, lost and reimposed (see Secondary Dormancy in Seeds).

Morphological dormancy occurs in undifferentiated embryos (which are often found in very small seeds <1 mm in length) and in differentiated linear or rudimentary embryos. The latter two types of embryo can have or develop physiological dormancy, thus increasing the complexity of the dormancy-breaking treatment. Morphological dormancy is often overcome by treatment at c. 15–30° C for many months to enable the embryo to grow. When combined with physiological dormancy, the necessary treatment may

Table 10.1. *Seed dormancy types (after Baskin and Baskin, 1998 (3))*

Type of dormancy	Features	Treatments to facilitate physiological response	Representative species
Physiological	Mostly permeable to water, but embryo is physiologically inhibited. Covering structures of the seed may play a role in preventing germination. Cannot germinate at any temperature or only over a very narrow range. Low permeability of embryo covers to oxygen. Covers prevent inhibitor leaching. Covers can also mechanically restrict embryo growth.		
1. Non-deep		Relatively short periods (1–12 weeks) of cold; longer periods of dry warmth, or wet warmth (weeks to months). Light. Chemical application, e.g., KNO_3 and gibberellin. Isolated embryos usually grow.	Common in weeds; garden flower species, some woody plants, etc.
2. Intermediate		Cold (up to 6 months), but dry storage may reduce this requirement. Gibberellin, but rarely on the intact dispersal unit ("seed"). Isolated embryos usually grow.	Some maple (*Acer*) species, beech, *Polygonum* sp., etc.
3. Deep		Cold (approximately 2 to 5 months). Even isolated embryos will not grow or produce abnormal seedlings. Gibberellin is ineffective.	Apple, some *Sorbus* and *Prunus*, etc.

Morphological	Differentiated linear or rudimentary embryos. Can have or develop physiological dormancy.	Optimum temperature for embryo growth approximately 15–30° C (1 week to many months). May respond to light.	Arum lilies, ginkgo, buttercups, palms, etc.
	Undifferentiated embryos, in dwarf or micro seeds usually less than 1 mm in length.	Seeds may respond to cold or warm stratification, and will need to associate with a plant or fungal associate for seedling growth.	Orchids, broomrapes, etc.
Morpho-physiological	Rudimentary or linear embryos.	Warm followed by cold, or vice versa, or only cold or warm. Conditions enable embryo growth, then physiological dormancy removal.	Arum lilies, poppies, buttercups, magnolias, etc.
Physical	Impermeability to water (of the seed or fruit coat). Rarely associated with physiological dormancy.	Physical disruption of the impermeable layer (e.g., surgical intervention).	Many legumes, Cistus, sumaks, etc.
Chemical	Inhibitors in the fruit or seed, potentially translocated towards the embryo.	Leaching in water, to remove abscisic acid? But many studies on seeds with physiological dormancy!	?
Mechanical	Presence of a hard, woody fruit wall.	May need surgical intervention, or chilling, indicating physical and physiological components, respectively!	Olives, Brazil nut, sumaks

involve only warm, or only cold or warm followed by cold, or vice versa. Seeds of poppies, arum lilies and buttercups fit this category.

Seed Dormancy Induction

Most studies have focussed on physiological dormancy in seeds with fully grown embryos. Seed dormancy can be imposed at two stages in the seeds' life: during development, in which case dormancy is referred to as "primary;" and/or after seed dispersal, so-called "secondary" dormancy (6, 29). Secondary dormancy is reversible, leading to the possibility of an annual dormancy cycle in the soil seed bank (see Secondary Dormancy in Seeds).

Development of Primary Dormancy and Desiccation Tolerance

By about halfway through development most seeds acquire the ability to germinate. Germination is not suppressed in the mangrove species *Avicennia marina* and in viviparous mutants, for example maize (VP mutants) or *Arabidopsis* (ABA-deficient or -unresponsive) (37). Some genotypes of crops are also prone to pre-harvest sprouting on the parent plant. In most other systems, though, germination of developing seeds is blocked by a combination of abscisic acid (ABA) and osmotic environment. Both of these factors inhibit, via sensitivity and concentration effects, water uptake and thus embryo growth. Their contribution to seed dormancy varies between species, having interactive effects in muskmelon (91), but their action(s) is(are) not necessarily interdependent. For example, both exogenously applied ABA and osmotic stress induce different genes involved in the induction of desiccation tolerance (57), a trait that develops later than the ability to regulate germination (38).

In seeds, the acquisition of desiccation tolerance is often perceived as being closely associated with the ability to enter the dormant state, or even to represent the same type of process. But the two are separate. For example, most but not all desiccation-sensitive seeds are non-dormant (8). Moreover, it is not the case that seeds must be dormant to become desiccation tolerant. Nonetheless, even partial drying can modulate the dormant state. For example, in the desiccation-sensitive seeds of the horse chestnut, partial desiccation of UK material to about 40% moisture content *ex planta* results in a reduction in the depth of dormancy (84).

Similarly, in bacteria the relationship between "dormancy" and desiccation tolerance can be close, but the two are not necessarily directly related. For example, the switch from a vegetative growing cell to a spore can be triggered by depletion of a readily metabolised form of carbon, nitrogen or phosphate. This strategy helps the bacteria to survive nutritionally unfavourable conditions. The spore core accumulates cations, is mineralised with calcium and/or hydrogen and is often chelated to dipicolinic (pyridine 2,6-dicarboxylic) acid (DPA). The consequence of these changes are reduced spore water content, undetectable metabolism, and the acquisition of a high tolerance of physical stresses, including extreme desiccation (52). There is probably a role for dipicolinic acid in the establishment and/or maintenance of dormancy in *Bacillus cereus* and other species of the same genus but not a direct role in resistance as DPA-negative heat-resistant revertants contain no DPA and low levels of calcium (25, 52). As seeds develop further, and their moisture content approaches or falls below about 50% (this is species dependent), their osmotic environment probably contributes more to germination suppression than ABA, since ABA levels usually decrease before the moisture content reaches the 50% level (38). However, for seeds maturing in fleshy fruits, chemical inhibitors probably continue to contribute to the induction/maintenance of dormancy (3).

After the vascular connections to the parent plant are severed, the seed starts to equilibrate with the natural environment and the seed moisture content falls further. At this stage of development, seeds have already become fully desiccation tolerant, and dormancy can either become deeper or shallower as drying progresses. For example, desiccation to <90% RH of isolated seeds of *Carica papaya* (papaya) results in physiological dormancy, perceived in this instance as a reduced competence for germination at a constant temperature of 26° C and a preference for alternating temperature or the requirement for stimulation by heat shock applied when the seeds have rehydrated (92). A few other species enter dormancy on drying and/or dry storage, including two conifer species (*Pseudotsuga menziesii* and *Picea sitchensis*) and the sand dune grass *Uniola paniculata* (see 92).

More commonly, as seeds reach moisture levels approaching air dryness, i.e., <75% RH or approximately <15% moisture content (27), primary dormancy decreases, in which case the seeds are said to have after-ripened (3, 6). Such conditions can also be duplicated in the laboratory and can be applied as an effective dormancy-breaking test/treatment (see Overcoming Primary Dormancy in Seeds and Other Systems).

The maturation environment has a major impact on dormancy. Withholding water from the plant can affect the seed dormancy level, which is lower in *Avena fatua* – wild oat (see 51). Thus, the proportion of wild oat seeds that can germinate without the application of a dormancy-breaking treatment is higher and/or the sensitivity of the seed population to dormancy-breaking treatment is increased when the plants have been water-stressed. The opposite is true of *Hordeum vulgare* – barley (see 51). Moreover, warmer temperatures during maturation usually reduce dormancy, as they do in *Avena fatua* – wild oat – (61), *Aesculus hippocastanum* – horse chestnut – (69) and many other species (3).

One of the key features in the development of desiccation tolerance in seeds is protection of DNA. During the development of dormancy, DNA replication is one of the earliest events to be curtailed, followed by RNA transcription and, later, protein synthesis. By the time desiccation-tolerant seeds are shed, the majority of nuclei have arrested in G_1 with a 2C DNA amount, i.e., the nuclear genome is diploid (9). Moreover, DNA replication and cell cycling cannot proceed in dormant seeds, but DNA repair can, and this is also the case in non-dormant wild oat embryos treated with ABA (18). However, the presence of 4C DNA (post-amplification of nuclear DNA but pre-mitosis) in embryos is not necessarily an indication of desiccation sensitivity. Rather, based on studies on "germinating" rye and oat embryos that had reduced desiccation tolerance and on recalcitrant seeds of *Avicennia marina*, it appears that increased desiccation sensitivity relates to a reduced competence for DNA repair when water is again made available (9, 58).

In developing *Bacillus subtilis* spores, the expression of specific low-molecular-weight acidic (SASP) proteins coincides with the loss of water from B-form DNA duplexes that results in the formation of a more stable A-form DNA (77). These α/β SASP proteins are synthesised only during sporulation and are triggered into action by low pH and dipicolinic acid (35). The proteins are released again during rehydration by proteases synthesised during dehydration. Apart from stiffening the nucleic acid, SASP protein binding to DNA reduces its reactivity with a variety of chemicals; in other words it increases stability of DNA. These proteins are degraded early in rehydration/germination and SASP-deficient mutants exhibit greater sensitivity of DNA to damage. However, DNA damage in dormant spores may be repaired during germination via special proteins. For example, the thyminyl–thymine adduct is involved in the repair of some of the lesions caused by UV irradiation.

In seeds, the development of stress tolerance is accompanied by the accumulation of many proteins, amongst which the late-embryogenesis-abundant (LEA) and heat-shock proteins (HSP) are important (38). But evidence for specific-function DNA-binding proteins in seeds is lacking (9).

Structural Features of Dormancy

In some species, the onset of physical dormancy coincides with the reduction of seed moisture content to relatively low levels resulting in hardening of the seed coat which prevents water uptake. Such "hardseededness" occurs in many legumes (*Fabaceae*) upon drying below about 11–14% moisture content (3, 51). Often, further desiccation is possible through a one-way valve in the seed coat. The coat then acts as a rigid barrier, protecting the seeds against physical stresses such as compaction in the soil (see Dry Storage).

A similar structure seems to exist in the bacterial endospore, which is unique in prokaryotic cells. The core is surrounded by three membranes, and the inner and outer coat are surrounded with rigid cell walls as a result of peptidoglycan accumulation. Such a structure may have evolved in pre-Cambrian times (c. 590 million years ago (MYA), long before angiosperms (seed-bearing flowering plants) evolved c. 120 MYA.

A second structural feature of seed dormancy is polymorphism, i.e., more than one form of seed in a species or even on the same plant. This is a relatively common phenomenon in some species (see 3). Whilst in no way being a rule, thinner-coated seeds tend to be less dormant than their thicker coated siblings, as in *Chenopodium album*. Morphological variation in composite seeds such as the daisy family may also translate into differential dormancy; for example, in *Senecio jacobaea* – the tansey ragwort (48) – ray (outer) achenes (a dry indehiscent one-seeded fruit) are heavier, smoother (no hairs) and more dormant than disc (inner) achenes.

Overcoming Primary Dormancy in Seeds and Other Systems

Once an organism enters a state of developmental arrest (dormant state), growth does not necessarily resume when optimum conditions occur. Rather, an activating stimulus is required, although this is not usually needed for additional growth (22). What are these activating stimuli and are there similarities between diverse organisms?

Chemical Effects

The consequence of applying dormancy-breaking treatments to any organism is an altered growth potential and water status of the tissue (embryo, bud, spore, etc.), which facilitate emergence.

In seeds, there are three main methods of overcoming dormancy: dry after-ripening at warm temperatures (for example, 10% moisture content seeds held at 40° C); moist chilling, in which fully hydrated seeds are held at 5° C for many weeks; and a combination of alternating temperature and light (white light is normally applied but the red light component is the most active via activation of phytochrome). What, if anything, links the three methods? One contender is the plant hormone abscisic acid (ABA). It has been hypothesised that ABA may be degraded during dry after-ripening of sunflower seeds (7, 29), leached during wet chilling of lettuce seeds, and catabolised or conjugated following (red) light treatment in lettuce (86). Red light also enhances gibberellins in lettuce seeds. However, there is no simple relationship between gibberellic acid (GA) content and germination. Rather, GA synthesis and/or increasing sensitivity are also likely to be involved in the transition from seed dormancy to germination (29–31).

In addition to GA, nitrate is commonly used in solution to help overcome dormancy in seeds (Table 10.2). However, nitrate is but one of a number of dormancy-breaking chemicals that are effective in a diverse group of organisms. Many of these are low-molecular-weight substances, such as weak acids, aldehydes, ketones and alcohols. Of particular interest are the organic acids that are efficacious at breaking dormancy (ending developmental arrest) in monera, protists, fungi, plants and animals. These include methanoic, ethanoic, propanoic, butanoic, pentanoic and hexanoic acid (Table 10.2). The pH of dormant plant tissues and seeds tends to be higher compared to the non-dormant state, suggesting that these acids could trigger dormancy loss via a change in intracellular pH (22). Changes in intracellular pH also affect dormancy induction/loss, *inter alia*, in *Xenopus* eggs, nematode larvae and brine shrimp cysts (see 22).

In seeds, ABA induces increased cellular pH and the expression of dormancy-associated genes, including the transcription factor VP1 and its homologues, which are involved in embryo maturation and developmental arrest. Dry after-ripening in wild oat seeds results in reduced transcription of the VP1 homologue (36), which has been implicated in restricting reserve lipid breakdown that normally occurs prior to germination (59) and repressing germination-specific amylase genes (89).

Temperature Effects

No convincing explanation is commonly accepted for the effects of temperature on dormancy release, although the growth inhibitor, ABA (see Chemical Effects), and promoter, gibberellin(s), are implicated.

Seeds of winter annuals tend to germinate in the autumn; the plants overwinter and then flower and set seed in the spring or early summer. To ensure germination in the autumn, the seeds must be exposed to high summer temperatures (i.e., reaching 25 to 35° C daily) for several months to remove dormancy (3). This type of response is described as after-ripening, and its rate is dependent on the treatment temperature and the seed moisture content. For temperature, it appears that the logarithm of the mean dormancy period is a negative linear function of temperature with a Q_{10} of around 3. This means that the rate of change in the dormancy state is three times faster at a

Table 10.2. *Dormancy-breaking chemicals common to four or five kingdoms (after Footitt and Cohn, 2001 (22))*

Chemical	Monera (endospore)	Protists (cysts)	Fungi (spores)	Higher plants (seeds)	Animals (*juvenile/#egg/†pupae)
Methanoic acid	*Bacillus megaterium*	—	*Phycomyces blakesleeanus*	*Oryza sativa*	*Asterias forbesii* # † * *Nematospiroides dubius* * *Strongylocentrotus purpuratus* #
Ethanoic acid	*Bacillus megaterium*	*Hartmannella rhysodes* *Schizopyrenus russellii*	*Hygrophorus russula* *Phycomyces blakesleeanus*	*Oryza sativa* *Avena fatua*	*Arbacia* sp. # *Asterias forbesii* # *Strongylocentrotus purpuratus* # *Thalassema mellita* #
Propanoic acid	*Bacillus megaterium*	—	*Phycomyces blakesleeanus* *Tricholoma favovirens*	*Oryza sativa*	*Asterias forbesii* # *Strongylocentrotus purpuratus* #
Butanoic acid	*Bacillus megaterium*	—	*Hygrophorus russula* *Phycomyces blakesleeanus* *Tricholoma favovirens*	*Oryza sativa* *Avena fatua*	*Arbacia* sp. # *Asterias forbesii* # *Strongylocentrotus purpuratus* #
Pentanoic acid	*Bacillus megaterium*	—	*Phycomyces blakesleeanus* *Hygrophorus russula*	*Oryza sativa*	*Asterias forbesii* # *Strongylocentrotus purpuratus* #

(cont.)

Table 10.2. (cont.)

Chemical	Monera (endospore)	Protists (cysts)	Fungi (spores)	Higher plants (seeds)	Animals (*juvenile/#egg/†pupae)
Hexanoic acid	*Bacillus megaterium*	—	*Phycomyces blakesleeanus*	*Oryza sativa*	*Asterias forbesii* # *Strongylocentrotus purpuratus* #
Butanol	*Bacillus megaterium*	—	*Neurospora tetrasperma* *Phycomyces blakesleeanus*	*Oryza sativa* *Avena fatua* *Avena sativa* *Echinochloa crus-galli*	*Asterias forbesii* # *Sarcophaga crassipalpis* †
Ammonia	*Bacillus cereus*	—	*Phycomyces blakesleeanus*	*Oryza sativa* *Avena fatua* *Berbera verna*	*Arbacia* sp. # *Ascaris suum* * *Haemonchus contortus* * *Nematospiroides dubius* * *Polynoe* sp. #
Nitrate	*Bacillus megaterium*	*Hartmannella rhysodes* *Schizopyrenus russellii*	—	*Avena fatua* *Amaranthus albus*	*Asterias forbesii* #

given temperature than at 10° C lower. In cereals, after-ripening is greatest at around 11–15% moisture (fresh mass basis), which is probably around 50% RH. Higher relative humidities of about 70% are less effective (see 51). Such treatments might involve the deactivation of ABA (see Chemical Effects). As dormancy loss occurs, the seeds first gain the ability to germinate at low temperatures (e.g., at a mean of <15° C) and then with additional dormancy loss also at higher temperatures (3).

Temperature treatments for the removal of physiological dormancy in hydrated tissues (seeds and buds) mainly involve long-term chilling at a constant cold temperature or daily temperature alternation accompanied by light treatment. The large variation in individual seed dormancy within a population of seeds can mean that the positive effects of temperature on dormancy loss via chilling can be spread over two orders of magnitude. For example, from <1 week to ~26 weeks is the chilling requirement in *Arum maculatum* (*Araceae*) seeds (67). An even wider variation in response (four orders of magnitude) can occur when light is the dormancy-breaking treatment (see e.g., 20).

Dormancy in temperate species enables the seeds to overcome inclement weather during winter, with the dormant state lessening as the period of chilling progresses. But by what mechanism does this happen? One possibility is that chilling increases the fluidity of membranes as a result of cold-activated oleoyl phosphatidyl choline desaturase (ODS) enhancement of 18:2 fatty acids. There is some evidence for this in sunflower seeds treated at 10° C (see 31). A similar situation appears to exist in apple buds, where the linoleic acid content of the cell membranes increases during chilling-induced dormancy relief (21). Growth resumption is facilitated by further desaturation of membranes under warm temperatures (21). The consequent increase in polyunsaturated fatty acids, which depress membrane melting points, could permit regulated metabolism at the lower temperature.

One of the most frequently postulated mechanisms for control of seed dormancy is a block in mobilisation of reserves – lipid, protein and sugar. Of particular interest recently is the observation that the temperature optimum for acid lipase is identical to the optimum of cold-mediated dormancy removal in apple seed embryos (46).

Fully hydrated seeds may also respond to temperature alternation as part of a dormancy-breaking treatment. Examples of such alternation are 30/20° C for tropical material and 20/10° C for temperate species. Light is usually applied in the warm phase, which usually lasts 12 hours each day. Papaya (*Carica papaya*) seeds freshly extracted from the fruit will germinate (albeit rather slowly) when incubated at ~26° C in the light (92). However, desiccation induces dormancy, thereby reducing the seeds' ability to germinate under these conditions. An alternating temperature of 33/19° C improves germination but not as much as applying a short pulse of heat shock, in this instance 36° C for about 4 hours. Lower germination levels result if the seeds are left constantly at 36° C. Dormant seeds of *Mallotus japonicus* similarly respond to a single high-temperature treatment at 32 to 40° C for several hours (90). As in other systems, the benefit of heat shock is ultimately expressed as a change in germinability/growth at a lower temperature, indicating that the influencing mechanism leads to an immediate cascade of events.

Other diverse systems also respond to a single, short-term (approximately) 10° C increase in temperature. Tuberculosis bacteria, which enter "dormancy" (stationary phase) when cultured at 37° C under oxygen-limiting conditions, can mount a typical heat-shock response when exposed to 46° C (34). Other single cell "systems" also respond to heat shock. For example, 15 minutes at 48° C activates the spores of the fungus *Phycomyces blakesleeanus* and allows them to germinate (72).

The studies on heat shock and dormancy-breaking chemicals point to the possibility of shared dormancy-breaking mechanisms in all kingdoms (22, 92). Heat-shocked tobacco cells produce an mRNA for a protein that binds to calmodulin (47). Calmodulin is sensitive to sub-molecular changes in Ca^{2+} and acts as a molecular switch to regulate proteins, including enzymes such as NAD kinase. This enzyme is thought to be involved in dormancy breakage by chilling in hazelnuts (74). Ca^{2+} loading is a feature of the dormant (and resistant) state in *Bacillus subtilis* (see Development of Primary Dormancy and Desiccation Tolerance). Thus, aspects of signal transduction associated with dormancy loss may have been conserved in diverse organisms.

Secondary Dormancy in Seeds

Primary dormancy serves to prevent precocious germination in the parent plant and soon after dispersal, facilitating temporally dispersed asynchronous germination. Secondary dormancy may develop when moist seeds are exposed to anaerobic conditions or during prolonged aerobic conditions that permit the loss of dormancy (e.g., chilling) but not the progression of germination. Thus, initially primary dormancy is relieved, but as the period of non-germination progresses the seeds enter secondary dormancy, making them harder to germinate and more likely to form some sort of soil seed bank. Periodic changes in secondary dormancy may explain seasonal emergence of weedy species, with temperature and possibly soil water potential the predominant controlling factors (31, 70).

For more temperate species, induction of secondary dormancy is fastest at around 20° C, whilst for warm temperate species, induction is fastest at low temperatures close to 5° C (see 51). Loss of secondary dormancy occurs by means similar to those described for primary dormancy (see Overcoming Primary Dormancy in Seeds and Other Systems). Light is also a modulator of secondary dormancy (and germination in cultivated land), with high intensities of white light and far-red light inhibiting germination. In contrast, low light and nitrate tend to reduce or delay such dormancy. Without light stimulation, buried seeds remain dormant if a relatively uniform daily temperature is maintained, but since seasonal temperatures vary, such seeds will cycle in and out of dormancy (see 51).

Thus any model that tries to explain the cycling of seeds in and out of dormancy needs to account for the potential effects of temperature, chemical stimulants and light. Hilhorst (31) has recently proposed a model involving alterations in membrane permeability and transition temperature (both related to phospholipid saturation), and associated protein conformation (affecting light and chemical receptors).

Overcoming Physical Dormancy in Seeds

Hard seeds are most frequently found in species adapted to survive in seasonally dry environments (3). Such environments also include species that protect their seeds in hard fruits, such as the cones of *Banksia* and some pines. For hard seeds to germinate, the physical barrier to water uptake must be removed, and in drier environments these barriers are most likely thermal shock associated with occasional and brief flash fires or daily solar gain, i.e., direct heating by the sun. In Mediterranean climates, the soil surface may reach 60° C (71). *Trifolium dubium* (*Fabaceae*) and *Geranium solanderi* (*Geraniaceae*) seed germination increased when soil temperatures were above 50° C for a few hours per day (3). Flash fires are likely to briefly (less than about 5 minutes) heat seeds in the soil seed bank to about 100° C. Species that readily recolonise land devastated by bush fires include *Acacia* (legumes; *Fabaceae*) in Australia and *Rhus ovata* (*Anacardiaceae*) in the Californian chaparral (64). Seeds of other species that become permeable after treatment at about 100° C are *Ceanothus megacarpus* and *C. sanguineus* (*Rhamnaceae*; wild relatives of the blue flowered shrub for British gardens), *Cistus albidus* and *C. monspeliensis* (*Cistaceae*) and *Spharolobium vimineum* (*Fabaceae*) (3). More than very short-term exposure to temperatures much higher than this tends to be lethal for most seeds.

Differential expansion and contraction of the seed coat layers or areas of the seed coat, as a result of heating and cooling, reveal fissures through which water can later penetrate. Fire-related dispersal strategies have evolved in some conifers, such as *Pinus serotina* and *Pinus contorta*. The seeds remain in the cones on the tree, forming an aerial seed bank, until the cone is cracked open by the heat (64). Viable seeds may be extracted from cones that are many decades old.

Exposure of dry seeds to ultra-low temperature, such as that associated with liquid nitrogen, also leads to the propagation of cracks in seed coats, as in legumes (68), again overcoming this type of "dormancy." For germination testing in the laboratory, though, surgical (file or scalpel) intervention to breach the barrier is most commonly used (3).

THE PERSISTENT STATE, RESISTANCE TO DISEASE AND HERBICIDES

Being able to control the precise switch from the dormant to the non-dormant state in seeds is of great value to agronomists. As a consequence, much effort has been directed towards elucidating the molecular basis of the dormant and non-dormant states. Clearly, possibilities already exist to manipulate (breed/select/transform) this trait. However, this raises the question of whether genotypes that have dormancy bred into them have greater tolerance to a range of biotic and abiotic stresses, thus leading to an enhanced risk of persistence. If this is the case, then opportunities might exist for the development of "weediness," i.e., first to establish on cleared ground. Whilst the application of herbicides can control the threat of such species to agricultural production, the release (intentional or unintentional) of herbicide-resistant transformed plants potentially poses a different threat to other plants, e.g., through genetic contamination via hybridisation with very closely related wild plants.

Persistence in the Dry and Wet State

The fecundity of higher plants means that the number of seeds produced each year is, generally, far in excess of that needed for next year's growth and establishment. As a consequence, ungerminated seed can become buried. But how long will such seeds persist in the natural environment? Persistence relates to a number of seed features and the environment into which they are dispersed.

Seed persistence did not evolve and does not operate in isolation, but forms a well adapted phenotype (83). The following seed features have an impact on potential persistence:

(a) the presence of physical dormancy;

(b) the depth of physiological dormancy of the seeds and the ability to regain dormancy (secondary dormancy);

(c) the inherent longevity of the seeds in the wet or dry state, depending on the natural environment; and

(d) defence against predation and pathogens.

Dry Storage

The only type of dormancy that plays a direct role in seed persistence is hardseededness or physical dormancy (see Overcoming Physical Dormancy in Seeds) when the seeds remain dry during storage in the soil. Such sealing in of dryness probably contributed to the exceptional longevity reported for seeds of *Nelumbo nucifera* – sacred lotus. These viable seeds, which were encased in the dry, hard fruit, were buried in a lake bed deposit in southern Manchuria, and radiocarbon dating indicated a survival period of around 600 years (79). A similar record exists for seed of *Canna compacta* (*Cannaceae*) found within a walnut shell rattle in Argentina. However, subsequent seedling growth was not completely normal (64). One seed of *Medicago polymorpha* recovered from an adobe building in California/northern Mexico germinated at an estimated age of 200 years. Interestingly, non-hard seeds of *Hordeum vulgare* (barley) and *Avena sativa* (oat) showed 90 and 81% germination, respectively, after 110 years of hermetic sealed storage in the Vienna herbarium at temperatures between 10 and 15° C and at about 3% moisture content (81). Other seed samples of the same two species survived (10–20% germination) storage at about 7% moisture content for 124 years in a Nuremberg building (2).

This evidence lends support to the predictions (based on modelling the effects of reducing moisture content and temperature on longevity) that dry seeds of many species could survive for centuries under conventional seed bank conditions of around 5% seed moisture content and −20° C (17). Bacteria appear to be even more successful survivors in the dry state (anhydrobiosis), as longevity (persistence) has been reliably measured at up to 10^5 years (see 62).

Once dry, seeds and spores have elevated levels of tolerance of high temperature. Dry heat survival of hard seeds of *Acacia melanoxylon* (*Fabaceae*) is about 10 minutes at 103° C, nearly all viability being lost by 100 minutes (66). By comparison, *Lolium perenne* (*Poaceae*; grasses) seeds are much more tolerant; 40% of the seeds germinate after 5 hours at 100° C (14). In contrast, dry heat survival of *Bacillus subtilis* spores at 100° C is about 9 days. Spore demineralisation, however, reduces resistance to dry heat, e.g., in *Bacillus stearothemophilus* (see 52).

Nonetheless, seeds in dry storage eventually accumulate injury. For example, in rye seeds an increase in the number of single- and double-stranded breaks in nuclear DNA occurs during dry storage (19). Only during subsequent rehydration can energy-requiring repair take place (18, 19). Priming or aeration-hydration treatments, at high water content insufficient for germination, can enhance longevity (a contributor to persistence) probably by activating DNA repair systems (63). In contrast, rehydration beyond the point of DNA replication tends to result in seeds that are intolerant of subsequent dehydration (9).

Similarly, dry heat (and radiation) damage in bacteria occurs through DNA, with greater sensitivity accompanying α/β-type SASP deficiency. As for seeds, DNA damage in dormant spores may be repaired during germination via special proteins. For example, the thyminyl–thymine adduct is involved in the repair of lesions caused by UV irradiation, (although not all UV damage relates to this adduct; see 52). Cellular proteins may also be damaged during storage, with aspartate-O-methyltransferase, methionine sulfoxide reductase and heat-shock proteins all implicated in recovery or protection. Spore demineralisation can also reduce resistance to dry heat, e.g., in *Bacillus stearothemophilus* (1). Finally, one particular form of Cyanobacteria – the Nostoc commune – has a marked capacity to tolerate anhydrobiosis, via numerous adaptations, including conspicuous extracellular glycan, synthesis of UV-absorbing pigments and maintenance of protein stability and structural integrity (see 52).

A gaseous environment has relatively little effect on dry seed storage longevity in a wide range of species, with modest benefits resulting from the use of nitrogen compared to air (64). Similarly, vacuum storage of seeds appears to offer little advantage over other sealed treatments, except for moderate benefits observed with, for example, American elm (*Ulmus americana*), various coniferous species, onion, cotton and sorghum (64). Not only dry seeds tolerate vacuum treatment. According to Horneck (32), up to 70% of bacterial and fungal spores survive short-term (10-day) exposure to a space vacuum, even without protection. When protection was present in the form of glucose, 80% of *B. subtilis* survived nearly 6 years of space vacuum (33). The inference is that microbes might survive interplanetary transfer. Support for this notion comes from the recovery of human oral bacterium, *Streptococcus mitis*, from a camera case left on the lunar surface for 2.5 years (50, 52), although secondary contamination probably cannot be ruled out completely.

Dry seeds are tolerant of numerous chemical assaults because of the protective nature of their covering structures (seed coat and fruit coat, depending on species). This is especially true of hardseeded legumes, and sulphuric acid treatment can be used to partly "digest" the coat of such seeds, thereby allowing the uptake of water as a prelude to germination (3). Bacterial spores too are generally significantly more resistant than growing cells to a wide variety of toxic chemicals (acids, bases, phenols, aldehydes and alkylating and oxidising agents). In this case, the proteinaceous spore coats protect the core from large molecules (lytic enzymes) and smaller molecules (including iodine and oxidising agents) perhaps by exclusion (impermeability) and/or cross-reaction (detoxification). However, the coats play no role, apparently, in spore resistance to alkylating agents. Generally, α/β-type SASP binding of DNA in spores makes them more resistant to chemical attack (52).

Subjecting seeds to irradiation is a common way of inducing damage to DNA (58) and accelerating the rate of viability loss through the accumulation of organic free radicals (64). However, the response is moisture-content dependent, with seeds in sorption zone II of water absorption, i.e., about 5–10% moisture content, being less susceptible to such treatment than both drier and wetter seeds (64). Bacterial spores are about 10–50 times more resistant than growing cells to UV radiation at 254 nm in water (76). Resistance probably relates to an altered UV photochemistry brought about by the saturation of spore DNA with α/β-type SASP resulting in the production of a unique thymine addict not present in cells. This resistance also applies in the dry state. In addition, there is evidence from spore coat–defective mutants of *B. subtilis* that the inner coat, in particular, plays some role in resistance to environmentally relevant UV wavelengths. Finally, there is a transient period of elevated UV resistance in germinating spores which possibly results from reduced photoreactivity of DNA as it passes through a transitional conformation (from A to B) (78).

Seeds suitable for banking readily tolerate desiccation to moisture contents in equilibrium with water potentials of about −250 MPa and transfer to liquid nitrogen temperature, at less than −150° C (65, 73). In contrast, only about 50% of spores of *Bacillus subtilis* and *Clostridium mangenoti* survived desiccation at 10^{-6} Pa (−10 MPa) and 77 Kelvin (−196° C) for 24 hours, whilst all *Escherichia coli* and *Halobacterium halobium* cells were killed (41, 52).

Wet Storage

We might assume that persistence in the wet state is simply related to dormancy, but the evidence does not support this assumption (3, 83). Initially dormant seeds of *Rumex crispus* soon become non-dormant in the soil but can nonetheless remain persistent. Alternatively, seeds shed in a deeply dormant state may not necessarily produce a persistent seed bank, e.g., *Heracleum sphondylium*. Dormancy, then, is neither a necessary nor a sufficient condition for persistence in soil (83). Where cycling of physiological dormancy is a feature of the seed response (see Secondary Dormancy in Seeds), a lack of suitable germination conditions, rather than dormancy *per se*, modulates persistence.

In seeds wet enough for both anabolic and catabolic metabolism, damage occurs with storage time. However, provided sufficient oxygen is available, such damage may be repaired (88). It should be noted that this is not to ignore the fact that wetland species can produce dense seed banks in waterlogged soils – presumably using adapted metabolism to survive (44).

In lettuce seeds longevity in the dry state is estimated to be about one order of magnitude longer than in the wet state. For example, seeds at 5 and 40% moisture content have a longevity (the time to lose two probits of viability, e.g., from 84 to 16%) at 35° C of about 1,000 and 100 days (73). In comparison, the wet heat survival at about 100° C for *Bacillus subtilis* spores is 20–30 minutes, which is three orders of magnitude shorter than for dry heat. Here, spore survival time was chosen as the period taken for a decimal reduction in viability, the D value (52).

When seeds of the North American plant *Rhus laurina* (Anacardiaceae) and two legume species were treated at 85° C to remove hardseededness (physical dormancy), seed quality (viability) was reduced considerably more during 100 minutes of wet heat (hot water) compared to dry-heat treatment (66). In this experiment, the seeds were initially hard and dry, but the wet-heat treatment facilitated the permeation of moisture, presumably leading to an accelerated rate of viability loss. Similarly, there is evidence that core water content is inversely related to bacterial spore wet-heat resistance (4).

Although DPA-deficient mutants of *Bacillus subtilis* are less heat resistant than the wild type, they exhibit no decrease in UV resistance. Moreover, other mutants have restored heat resistance even though DPA is not restored (see 52). Wet-heat resistance by spores is also a function of at least two other factors – SASP content and sporulation temperature. Spores prepared at higher temperatures are generally more resistant than spores prepared at lower temperatures (12). This treatment probably reduces core water content, but other factors such as protein stability are also important.

Of course, seed survival in the natural environment is likely to be a function of both temperature and moisture content, and both of these will vary over time in relation to the seasons. Thus, with the exception of hardseeds, most seeds in the soil seed bank will probably be in relative equilibrium with the surrounding conditions. Evidence of how long seeds can last under such variable conditions comes from two long-term experiments. Firstly, in 1879, W. J. Beal buried seeds of 20 species in open glass bottles containing moist sand, and 100 years later it was found that a few seeds of three species were capable of germination (39). In a separate experiment, J. W. T. Duvel buried seeds of 107 species of which 36 species survived for 39 years (85). Clearly then, the potential for seed tolerance of burial is on the order of decades up to a century in cool temperate environments.

Longevity of seeds in the wet state is obviously influenced by many features, such as repair efficiency and ease of cycling into and out of dormancy, perhaps representing an intrinsic property of the species (64). On the other hand, chemical defence is dependent on the composition of the seed. Similarly, bacteria-spore resistance appears to have two components (intrinsic and compositional), such that psychrophilic bacteria are much less heat tolerant than thermophilic bacteria (see 25).

Persistence can be used as a measure of weediness, but the contributory factors vary between species. In annual mercury (*Mercurialis annua*), a weed of arable land, weediness is due to a long primary dormancy in the seeds, seed longevity in the soil and a high seed production (40). In contrast, the persistence of many forage legumes is typically based on their vegetative regeneration, even though seed regeneration can also be involved (82). Overall, the strongest co-correlant of seed persistence is seed size – small seeds are more persistent (16, 44).

Little is known about how spore-forming bacteria grow and survive under diverse natural conditions. However, there is evidence that dormant *Bacillus* spp. spores from soils of the Sonoran desert are significantly more UV resistant than the laboratory strain *Bacillus subtilis* 168 (53).

Depletion Rate in the Soil

Persistence in the soil also depends on the depletion rate, as affected by, *inter alia*, seed ageing, predation, germination (loss of dormancy) and germination without emergence (i.e., death whilst in the soil or "fatal" germination). For example, for propanil-resistant *Echinocloa colona* (syn. *E. colonum*), seed longevity in rice fields is affected more by seed decay than by *in situ* germination (11).

Various models have been used to describe the empirical data and to predict future trends. Perhaps of greatest biological relevance is a negative exponential model for viable seed population density, meaning that the probability of depletion is constant each year and that persistent seed banks have a constant half-life (51). Depletion rate varies with species and increases with increasing frequency of cultivation. For undisturbed soils, it is possible that depletion of the vast majority of seeds could take close to 30 years (see 51).

Defence Mechanisms

Persistence in the soil depends on avoiding germination under inappropriate conditions. However, even among buried seeds that avoid immediate germination longevity is tremendously variable – from months to decades. In such circumstances, protection against predation and pathogen attack are important factors in survival. The predation level for buried seeds is relatively low compared to that for seeds on the soil surface, and thus the front line of defence is chemical, probably against microbial attack. In a detailed study of 81 species, Hendry *et al.* (28) showed that there were significantly higher levels of ortho-dihydroxyphenol in seeds that formed persistent seed banks. These chemicals have both bactericidal and fungicidal activity and can be toxic to some herbivores (see 28). However, some persistent seeds have low levels of ortho-dihydroxyphenol, and whilst tannins also may play a role in persistence, our understanding of the chemical basis of persistence in the soil is rather limited (83).

Herbicide Resistance

There are understandable concerns about the selective pressure that herbicide application might have on the development of "super" weeds with high chemical tolerance and long persistence. Clearly, any linkage between changes in seed dormancy and herbicide resistance traits has potential ecological and agronomic implications especially if associated with transgenic material.

Arabidopsis thaliana mutant plants that tolerate the bleaching effect of norflurazon (nfz), a carotenoid biosynthesis inhibitor, do so either by increased endogenous levels of ABA or increased sensitivity to ABA. However, it is not clear if changing seed responsivity to this dormancy-related compound impacted their dormancy status (80).

One of the better studied weed systems in the United States is red rice (*Oryza sativa*), the same species as cultivated rice. Red rice contributes to processing problems and yield losses that amount to about $50 million per year in the southern United States (15). Competition, large seed production, the tendency of panicles to shatter at maturity and seed dormancy are identified as important factors in its weed status. Promising results in controlling red rice have been achieved with acetolactate synthase–inhibiting herbicide, for which resistant rice lines have been bred by traditional methods. The genetic and agronomic consequences (vegetative and reproductive traits) of transferring glufosinate (Liberty TM) herbicide resistance from two transgenic rice lines to the four biotypes of the noxious weed red rice have been evaluated under field conditions. Seed dormancy and seed production were not significantly different among transgenic and non-transgenic populations. Populations segregating for glufosinate resistance responded in a location-specific manner with respect to life history and fecundity traits (55).

In a study evaluating resistance classes of rigid ryegrass (*Lolium rigidum*) to aryloxyphenoxyproprionate (AOPP) and sulfonylurea (SU), no significant differences were detected among resistant and susceptible classes with respect to seed dormancy, relative growth rate and phenological development (24). Similarly, seed from both acetolactate (ALS)-resistant and -susceptible *Sisymbrium orientale* plants growing in South Australia were found to exhibit strong primary dormancy (10). So, herbicide-resistant plant lines do not necessarily exhibit an enhanced level of seed dormancy compared to control plants.

Seeds of diclofop-methyl–resistant Italian ryegrass (*Lolium multiflorum*) with a greater degree of dormancy were more likely to be herbicide resistant than those with a lesser degree of dormancy (23). Similarly, it was found in wild oat (*Avena fatua*) – a common weed of arable land, especially wheat fields – that accessions possessing high seed dormancy, herbicide resistance (to diclofop, difenzoquat, flamprop and MSMA) and seed production were more likely to become weedy (49).

In herbicide-resistant Italian ryegrass (*Lolium multiflorum*), the timing of field tillage/cultivation and herbicide use have been identified as important selective forces for increasing seed dormancy and thus for changing seedling emergence patterns; experimental plots had three times as many viable, resistant seeds as susceptible seeds (23). Land management is also an important factor in the evolution of herbicide resistance in *Avena fatua*. In a study over 18 years, it was found that continuous cultivation and tri-allate application enhanced resistance to the herbicide in up to 14% of the seeds screened. However, fallow in the rotation slowed the rate of evolution of resistance, and this may be associated with dormancy-induced delays in growth (5).

Local environmental conditions can also affect the impact of herbicides on wild oat. In a study in Canada, phytotoxicity of imazamethabenz and fenoxaprop was found in four genetically distinct lines to be influenced adversely by short- to long-term water stress. These variable responses to herbicide among populations (lines) did not appear, though, to be directly related to their intrinsic drought tolerance, growth habit or to their level of seed dormancy (93). Nonetheless, it is likely that only by understanding more about the natural biology of the species – seed dormancy, germination, life cycle, longevity and phenology – will it be possible to improve the management of species, thereby lowering the reliance on herbicides. An example of where such an approach is being taken is *Cyperus difformis*, a weed of rice fields in Australia. Control of this species has relied heavily on the application of the herbicide bensulfuron-methyl and resistant biotypes are now known to exist in California (75).

Overall, the evidence so far suggests that there is no simple relationship between herbicide resistance, dormancy, persistence and weediness.

CONCLUSION

In the dormant state, seeds and bacterial spores have no detectable metabolism and a high tolerance of physical stresses (3, 52). Clearly the strategy is highly successful as seeds and bacteria can be found in a wide range of habitats/environments on the earth, including those that are removed spatially from their origin (e.g., strictly thermophilic bacteria recovered from cold lake sediments). Some features of dispersal and persistence are shared by seeds and bacterial spores, as are some aspects of dormancy loss (e.g., chemical triggers and heat shock), and these have undoubtedly contributed to their respective evolutionary successes. Whilst it is tempting to focus our studies on model systems, especially exploiting the power of modern molecular biology, greater attention in the future should be given to the responses of diverse, wild-type organisms as these could help to further elucidate mechanisms of dormancy, persistence and resilience to abiotic and biotic stresses.

REFERENCES

1. Alderton, G., and Snell, H. 1969. Chemical states of bacterial spores: Dry heat resistance. Appl. Microbiol. **17**:745–749.
2. Aufhammer, G., and Simon, U. 1957. Die samen landwirtschaftlicher Kulturpflanzen im Grundstein des ehemaligen Nürnberger Stadttheaters unde ihre Keimfähigheit. Z. Acker-Pflanzenb. **103**:454–472.
3. Baskin, C. C., and Baskin, J. M. 1998. *Seeds. Ecology, Biogeography, and Evolution of Dormancy and Germination*. Academic Press, San Diego.
4. Beaman, T. C., and Gerhardt, P. 1986. Heat resistance of bacterial spores correlated with protoplast dehydration, mineralisation, and thermal adaptation. Appl. Environ. Microbiol. **52**:1242–1246.
5. Beckie, H. J., and Jana, S. 2000. Selecting for triallate resistance in wild oat. Can. J. Plant Sci. **80**:665–667.
6. Bewley, J. D., and Black, M. 1994. *Seeds. Physiology of Development and Germination*, 2nd ed., Plenum Press, New York.

PRITCHARD AND TOOROP

7. Bianco, J., Garello, G., and Le Page-Degivry, M. T. 1994. Release of dormancy in sunflower embryos by dry storage: Involvement of gibberellins and abscisic acid. Seed Sci. Res. 4:57–62.

8. Black, M., and Pritchard, H. W. (Eds.). 2002. *Desiccation and Survival in Plants. Drying without Dying.* CAB International, Wallingford, UK.

9. Boubriak, I., Kargiolaki, H., Lyne, L., and Osborne, D. J. 1997. The requirement for DNA repair in desiccation tolerance of germinating embryos. Seed Sci. Res. 7:97–105.

10. Boutsalis, P., and Powles, S. B. 1998. Seedbank characteristics of herbicide-resistant and susceptible *Sisymbrium orientale.* Weed Res. 38:389–395.

11. Chaves, L., Valverde, B. E., and Garita, I. 1997. Effects of time and soil depth on the persistence of *Echinochloa colona* seeds. Manejo Integrado de Plagas 45:18–24.

12. Condon, S., Bayarte, M., and Sala, F. J. 1992. Influence of the sporulation temperature upon the heat resistance of *Bacillus subtilis.* J. Appl. Bacteriol. 73:251–256.

13. Cohn, M. A. 1996. Operational and philosophical decisions in seed dormancy research. Seed Sci. Res. 6:147–153.

14. Crosier, W. 1956. Longevity of seeds exposed to dry heat. Proc. Assoc. Official Seed Analysts 46:72–74.

15. Croughan, T. P., Utomo, H. S., Sanders, D. E., and Braverman, M. P. 1996. Herbicide-resistant rice offers potential solution to red rice problem. LA Agr. 39:10–12.

16. Dalling, J. W., Hubbell, S. P., and Silvera, K. 1998 Seed dispersal, seedling establishment and gap partitioning among tropical pioneer trees. J. Ecol. 86:674–689.

17. Dickie, J. B., Ellis, R. H., Kraak, H. L., Ryder, K., and Tompsett, P. B. 1990. Temperature and seed storage longevity. Ann. Bot. 65:197–204.

18. Elder, R. H., and Osborne, D. J. 1993. Function of DNA synthesis and DNA repair in the survival of embryos during early germination and dormancy. Seed Sci. Res. 3:43–53.

19. Elder, R. H., Dell'Aquila, A., Mezzina, M., Sarasin, A., and Osborne, D. J. 1987. DNA ligase in repair and replication in the embryos of rye, *Secale cereale.* Mutat. Res. 181:61–71.

20. Ellis, R. H., Hong, T. D., and Roberts, E. H. 1989. Quantal response of seed germination in seven genera of Cruciferae to white light of varying photon flux density and photoperiod. Ann. Bot. 63:145–158.

21. Erez, A. 2000. Bud dormancy: A suggestion for the control mechanism and its evolution. In *Dormancy in Plants. From Whole Plant Behaviour to Cellular Control,* Viémont, J.-D., and Crabbé, J. (Eds.), pp. 23–33. CAB International, Wallingford, UK.

22. Footitt, S., and Cohn, M. A. 2001. Developmental arrest: From sea urchins to seeds. Seed Sci. Res. 11:3–16.

23. Ghersa, C. M. M. A., Brewer, T. G., and Roush, M. L. 1994. Selection pressure for diclofop-methyl resistance and germination time of Italian ryegrass. Agron. J. 86:823–828.

24. Gill, G. S., Cousens, R. D., and Allan, M. R. 1996. Germination, growth, and development of herbicide resistant and susceptible populations of rigid ryegrass (Lolium rigidum). Weed Sci. 44:252–256.

25. Gould, G. W. 1986. Water and the survival of bacterial spores. In Membranes, Metabolism, and Dry Organisms, Leopold, A. C. (Ed.), pp. 143–156. Comstock Publishing Associates, Ithaca, NY.

26. Harper, J. L. 1957. The ecological significance of dormancy and its importance in weed control. Int. Congr. Plant Protection 4:415–420.

27. Hay, F. R. 1997. The Development of Seed Longevity in Wild Plant Species. Ph.D. thesis, University of London, UK.

28. Hendry, G. A. F., Thompson, K., Moss, C. J., Edwards, E., and Thorpe, P. C. 1994. Seed persistence: A correlation between seed longevity in the soil and ortho-dihydroxyphenol concentration. Funct. Ecol. 8:658–664.

29. Hilhorst, H. W. M. 1995. A critical update on seed dormancy. I. Primary dormancy. Seed Sci. Res. 7:221–223.

30. Hilhorst, H. W. M. 1997. Seed dormancy. Seed Sci. Res. 7:221–223.

31. Hilhorst, H. W. M. 1998. The regulation of secondary dormancy. The membrane hypothesis revisited. Seed Sci. Res. 8:77–90.

32. Horneck, G. 1993. Responses of Bacillus subtilis spores to the space environment: Results from experiments in space. Orig. Life Evol. Biosph. 23:37–52.

33. Horneck, G., Bücker, H., and Reitz, G. 1994. Long-term survival of bacterial spores in space. Adv. Space Res. 14:41–45.

34. Hu, Y. M., Butcher, P. D., Sole, K., Mitchison, D. A., and Coates, A. R. M. 1998. Protein synthesis is shutdown in dormancy Mycobacterium tuberculosis and is reversed by oxygen and heat shock. FEMS Microbiol. Lett. 158:139–145.

35. Illades-Aguiar, B., and Setlow, P. 1994. Autoprocessing of the protease that degrades small, acid-soluble proteins in spores of Bacillus species is triggered by low pH, dehydration, and dipicolinic acid. J. Bacteriol. 176:7032–7037.

36. Jones, H. D., Peters, N. C. B., Holdsworth, M. J. 1997. Genotype and environment interact to control dormancy and differential expression of the VIVIPAROUS 1 homologue in embryos of Avena fatua. Plant J. 12:911–920.

37. Karssen, C. M., Brinkhorst van der Swan, D. L. C., Breekland, A. E., and Koorneef, M. 1983. Induction of dormancy during seed development by endogenous abscisic acid: Studies on abscisic acid deficient genotypes of Arabidopsis thaliana (L.) Heynh. Planta 157:158–165.

38. Kermode, A. R., and Finch-Savage, W. E. 2002. Desiccation sensitivity in orthodox and recalcitrant seeds in relation to development. In *Desiccation and Survival in Plants*, Black, M., and Pritchard, H. W. (Eds.), pp. 149–184. CAB International, Wallingford, UK.

39. Kivilaan, A., and Bandurski, R. S. 1981. The one hundred-year period for Dr. Beal's seed viability experiment. Am. J. Bot. **68**:1290–1292.

40. Kohout, V., and Hamouz, P. 2000. Annual mercury (*Mercurialis annua* L.): Reasons for expansion on arable land. Zeitschrift Pflanzenkrankheiten Pflanzenschutz Special Issue. **17**:143–144.

41. Koike, J., Oshima, T., Koike, K. A., Taguchi, H., Tanaka, R., Nishimura, K., and Miyaji, M. 1992. Survival rates of some terrestrial microorganisms under simulated space conditions. Adv. Space Res. **12**:271–274.

42. Lang, A. G. (Ed.). 1996. *Plant Dormancy. Physiology, Biochemistry and Molecular Biology.* CAB International, Wallingford, UK.

43. Lang, A. G., Early, J. D., Martin, G. C., and Darnell, R. L. 1987. Endo-, para- and ecodormancy: Physiological terminology and classification for dormancy research. Hort. Sci. **22**:371–377.

44. Leck, M. A. 1989. Wetland seed banks. In *Ecology of Soil Seed Banks*, Leck, M. A., Parker, V. T., and Simpson, R. L. (Eds.), pp. 283–305. Academic Press, San Diego.

45. Leopold, A. C. (Ed.). 1986. *Membranes, Metabolism, and Dry Organisms.* Comstock Publishing Associates, Ithaca, NY.

46. Lewak, S., Bogatek, R., and Zarska-Maciejewska, B. 2000. Sugar metabolism in apple embryos. In *Dormancy in Plants. From Whole Plant Behaviour to Cellular Control*, Viémont, J.-D., and Crabbé, J. (Eds.), pp. 47–55. CAB International, Wallingford, UK.

47. Lu, Y.-T., Dharmasiri, M. A. N., and Harrington, H. M. 1995. Characterisation of a cDNA encoding a novel protein that binds to calmodulin. Plant Physiol. **108**:1197–1202.

48. McEvoy, P. B. 1984. Dormancy and dispersal in dimorphic achenes of tansey ragwort, *Senecio jabobeae* L.(Compositae). Oecologia (Berlin) **61**:160–168.

49. Miller, S. D., Nalewaja, J. D., and Mulder, C. E. G. 1982. Morphological and physiological variation in wild oat *Avena fatua*. Agron. J. **74**:771–775.

50. Mitchell, F. J., and Ellis, W. L. 1972. Surveyor 3: Bacterium isolated from lunar-retrieved television camera. In *Analysis of Surveyor 3 Material and Photographs Returned by Apollo 12*, pp. 239–251. U.S. Government Scientific and Technical Information Office, Washington, D.C.

51. Murdoch, A. J., and Ellis, R. H. 2000. Dormancy, viability and longevity. In *Seeds: The Ecology of Regeneration in Plant Communities*, Fenner, M. (Ed.), pp. 183–214. CAB International, Wallingford, UK.

52. Nicolson, W. L., Munakata, N., Horneck, G., Melosh, H. J., and Setlow, P. 2000. Resistance of *Bacillus* endospores to extreme terrestrial and exterrestrial environments. Microbiol. Molec. Biol. Rev. **64**:548–572.

53. Nicolson, W. L., and Law, J. F. 1999. Method of purification of bacterial endospores from soils: UV resistance of natural Sonoran desert soil populations of *Bacillus* spp. with reference to *B. subtilis* strain 168. J. Microbiol. Meth. **35**:13–21.

54. Nikolaeva, M. G. 1977. Factors controlling the seed dormancy pattern. In *The Physiology and Biochemistry of Seed Dormancy and Germination*, Khan, A. A. (Ed.), pp. 51–74. North Holland Publishing Co., Amsterdam.

55. Oard, J., Cohn, M. A., Linscombe, S., Gealy, D., and Gravois, K. 2000. Field evaluation of seed production, shattering, and dormancy in hybrid populations of transgenic rice (*Oryza sativa*) and the weed, red rice (*Oryza sativa*). Plant Sci. **157**:13–22.

56. Okubo, H. 2000. Growth cycle and dormancy in plants. In *Dormancy in Plants. From Whole Plant Behaviour to Cellular Control*, Viémont, J.-D., and Crabbé, J. (Eds.), pp. 1–22. CAB International, Wallingford, UK.

57. Ooms, J. J. J., van der Veen, R., and Karssen, C. M. 1994. Abscisic acid and osmotic stress or slow drying independently induce desiccation tolerance in mutant seeds of *Arabidopsis thaliana*. Physiol. Plant. **92**:506–510.

58. Osborne, D. J., Boubriak, I., and Leprince, O. 2002. Rehydration of dried systems: Membranes and the nuclear genome. In *Desiccation and Survival in Plants*, Black, M., and Pritchard, H. W. (Eds.), pp. 343–364. CAB International, Wallingford, UK.

59. Paek, N. C., Lee, B.-M., Bai, D. G., and Smith, J. D. 1998. Inhibition of germination gene expression by Viviparous-1 and ABA during maize kernel development. Mol. Cells **8**:336–342.

60. Pammenter, N. W., and Berjak, P. 1999. A review of recalcitrant seed physiology in relation to desiccation-tolerance mechanisms. Seed Sci. Res. **9**: 13–37.

61. Peters, N. C. B. 1982. The dormancy of wild oat seed (*Avena fatua* L.) from plant grown under various temperature and soil moisture conditions. Weed Res. **22**:205–212.

62. Potts, M. 1994. Desiccation tolerance of prokaryotes. Microbiol. Rev. **58**:755–805.

63. Powell, A. A., Yule, L. J., Jing, H.-C., Groot, S. P. C., Bino, R. J., and Pritchard, H. W. 2000. The influence of aerated hydration seed treatment on seed longevity as assessed by the viability equations. J. Exp. Bot. **51**:2031–2043.

64. Priestley, D. A. 1986. *Seed Aging. Implications for Seed Storage and Persistence in the Soil*. Comstock Publishing Associates, Ithaca, NY.

PRITCHARD AND TOOROP

65. Pritchard, H. W. 1995. Cryopreservation of seeds. In *Methods in Molecular Biology, Vol. 38: Cryopreservation and Freeze-Drying Protocols,* Day, J. G., and McLellan, M. R. (Eds.), pp. 133–144. Humana Press Inc., Totowa, NJ.

66. Pritchard, H. W., Espinosa, P., and Culshaw, C. 1996. Hard seed impermeability. In *International Plant Genetic Resources Institute Annual Report 1995,* p. 61. IPGRI, Rome.

67. Pritchard, H. W., Wood, J. A., and Manger, K. R. 1993. Influence of temperature on seed germination and the nutritional requirements for embryo growth in *Arum maculatum* L. New Phytol. **123**:801–809.

68. Pritchard, H. W., Manger, K. R., and Prendergast, F. G. 1988. Changes in *Trifolium arvense* seed quality following alternating temperature treatment using liquid nitrogen. Ann. Bot. **62**:1–11.

69. Pritchard, H. W., Steadman, K. J., Nash, J., and Jones, C. 1999. Kinetics of dormancy release and the high temperature response in *Aesculus hippocastanum* seeds. J. Exp. Bot. **50**:1507–1514.

70. Probert, R. J. 2000. The role of temperature in the regulation of seed dormancy and germination. In *Seeds: The Ecology of Regeneration in Plant Communities,* Fenner, M. (Ed.), pp. 261–292. CAB International, Wallingford, UK.

71. Quinlivan, B. J. 1971. Seed coat impermeability in legumes. J. Aus. I. Agr. Sci. **37**:283–295.

72. Rivero, E., and Cerdá-Olmedo, E. 1987. Spore activation by acetate, propionate and heat in *Phycomyces* mutants. Mol. General Genetics **209**:149–153.

73. Roberts, E. H., and Ellis, R. H. 1989. Water and seed survival. Ann. Bot. **63**:39–52.

74. Ross, J. D. 1996. Dormancy breakage by chilling: Phytochrome, calcium and calmodulin. In *Plant Dormancy: Physiology, Biochemistry and Molecular Biology,* Lang, G. A. (Ed.), pp. 157–169. CAB International, Wallingford, UK.

75. Sanders, B. A. 1994. The life cycle and ecology of *Cyperus difformis* (rice weed) in temperate Australia: A review. Aus. J. Exp. Agr. **34**:1031–1038.

76. Setlow, P. 1988. Resistance of bacterial spores to ultraviolet light. Comments Mol. Cell. Bioph. **5**:253–264.

77. Setlow, P. 1992. DNA in dormant spores of *Bacillus* species is an A-like conformation. Mol. Microbiol. **6**:563–567.

78. Setlow, B., and Setlow, P. 1988. Absence of transient elevated UV resistance during germination of *Bacillus subtilis* spores lacking small, acid-soluble spore proteins α and β. J. Bacteriol. **170**:2858–2859.

79. Shen-Miller, J., Schopf, J. W., and Berger, R. 1983. Germination of *ca.* 700 year-old lotus seed from China: Evidence of exceptional longevity and seed viability. Am. J. Bot. **70**:78.

80. Soldatova, O. P., Ezhova, T. A., Ondar, U. N., Gostimskii, S. A., Konrad, U., and Artsaenko, O. 1996. Mutants of *Arabidopsis thaliana* (L.) Heynh. tolerant

of the carotenoid biosynthesis inhibitor norflurazon. Geneticka (Moskva) 32:956–961.

81. Steiner, A. M., and Ruckenbauer, P. 1995. Germination of 110-year-old cereal and weed seeds, the Vienna sample of 1877. Verification of effective ultra-dry storage at ambient temperature. Seed Sci. Res. 5:195–199.

82. Sulas, L., Franca, A., and Caredda, S. 2000. Persistence and regeneration mechanisms in forage legumes. Cahiers Options Mediterraneennes 45:331–342.

83. Thompson, K. 2000. The functional ecology of soil seed banks. In *Seeds: The Ecology of Regeneration in Plant Communities*, Fenner, M. (Ed.), pp. 215–235. CAB International, Wallingford, UK.

84. Tompsett, P. B., and Pritchard, H. W. 1998. The effect of chilling and moisture status on the germination, desiccation tolerance and longevity of *Aesculus hippocastanum* L. seeds. Ann. Bot. 82:249–261.

85. Toole, E. H., and Brown, E. 1946. Final results of the Duvel buried seed experiment. J. Agric. Res. 72:201–210.

86. Toyomasu, T., Yamane, H., Murofushi, N., and Inoue, Y. 1994. Effects of exogenously applied gibberellin and red light on the endogenous levels of abscisic acid in photoblastic lettuce seeds. Plant Cell Physiol. 35:127–129.

87. Viémont, J.-D., and Crabbé, J. (Eds.) 2000. *Dormancy in Plants. From Whole Plant Behaviour to Cellular Control.* CAB International, Wallingford, UK.

88. Villiers, T. A. 1974. Seed aging: Chromosome stability and extended viability of seeds stored fully imbibed. Plant Physiol. 53:875–878.

89. Walker-Simmonds, M. K. 2000. Recent advances in ABA-regulated gene expression in cereal seeds: Evidence for regulation by PKABA1 protein kinase. In *Seed Biology. Advances and Applications*, Black, M., Bradford, K. J., Vazquez-Ramos, J. (Eds.), pp. 271–276. CAB International, Wallingford, UK.

90. Washitani, I., and Takenaka, A. 1987. Gap detecting mechanism in the seed germination of *Mallotus japonicus* (Thunb.) Muell. Arg., a common pioneer tree of secondary succession in temperate Japan. Ecol. Res. 2:191–201.

91. Welbaum, G. E., Tissaoui, T., and Bradford, K. J. 1990. Water relations of seed development and germination in muskmelon (*Cucumis melo* L.). III Sensitivity of germination to water potential and abscisic acid during development. Plant Physiol. 92:1029–1037.

92. Wood, C. B., Pritchard, H. W., and Amritphale, D. A. 2000. Desiccation-induced dormancy in papaya (*Carica papaya* L.) seeds is alleviated by heat shock. Seed Sci. Res. 10:135–145.

93. Xie, H. S., Hsiao, A. I., and Quick, W. A. 1993. Influence of water deficit on the phytotoxicity of imazamethabenz and fenoxaprop among five wild oat populations. Environ. Exp. Bot. 33:283–291.

Index

INDEX